"Provides a wealth of information, both biographical and scientific . . . will help the reader to a better grasp of scientific thought. . . . Supporters of any theory will come away enlightened and better able to debate after reading this thorough work."
—*Anniston Star*

"Solid and quite readable . . . admirably fulfills the nonspecialist's needs . . . clarifies the meaning of Darwin's theory."
—*Kirkus Reviews*

"Excellent . . . illuminating."
—*Virginian-Pilot*

"Makes good use of the large body of information now in print to fashion a nicely balanced account of Darwin's personal and professional lives."
—*Science*

"A fascinating book."
—*Roanoke Times*

"The sections tracing development of concepts concerning evolution and related religious ideas are well done."
—*Publishers Weekly*

MICHAEL WHITE was a director of studies at d'Overbroeck's College, Oxford, and is now a professional writer and journalist. He is the author of several books, including the bestselling *Stephen Hawking: A Life in Science* and *Einstein: A Life in Science* (both with John Gribbin), a biography of Isaac Asimov, *The Unauthorized Life*, and *Breakthrough*, a book about the hunt for the breast cancer gene. He lives in London with his wife. **JOHN GRIBBIN** received his Ph.D. in astrophysics from Cambridge University before becoming a full-time science writer. He is the author of more than fifty books, including the bestselling *In Search of Schrödinger's Cat* and *In the Beginning*. He lives in East Sussex, England.

DARWIN

A LIFE IN SCIENCE

Michael White
and John Gribbin

A PLUME BOOK

PLUME
Published by the Penguin Group
Penguin Books USA Inc., 375 Hudson Street, New York, New York 10014, U.S.A.
Penguin Books Ltd, 27 Wrights Lane, London W8 5TZ, England
Penguin Books Australia Ltd, Ringwood, Victoria, Australia
Penguin Books Canada Ltd, 10 Alcorn Avenue, Toronto, Ontario, Canada M4V 3B2
Penguin Books (N.Z.) Ltd, 182–190 Wairau Road, Auckland 10, New Zealand

Penguin Books Ltd, Registered Offices: Harmondsworth, Middlesex, England

Published by Plume, an imprint of Dutton Signet,
a division of Penguin Books USA Inc.
Previously published in a Dutton edition.
First published in Great Britain by Simon & Schuster Ltd.

First Plume Printing, April, 1997
10 9 8 7 6 5 4 3 2 1

℗ REGISTERED TRADEMARK—MARCA REGISTRADA

The Library of Congress has catalogued the Dutton edition as follows:
White, Michael.
Darwin : a life in science / Michael White and John Gribbin.
p.   cm.
Includes bibliographical references and index.
ISBN 0-525-94002-2 (hc.)
ISBN 0-452-27552-0 (pbk.)
1. Darwin, Charles, 1809–1882.   2. Naturalists—England—
Biography.   I. Gribbin, John R.   II. Title.
QH31.D2W495   1996
575'.0092—dc20
[B]
95–24567
CIP

Printed in the United States of America

BOOKS ARE AVAILABLE AT QUANTITY DISCOUNTS WHEN USED TO PROMOTE PRODUCTS
OR SERVICES. FOR INFORMATION PLEASE WRITE TO PREMIUM MARKETING DIVISION,
PENGUIN BOOKS USA INC., 375 HUDSON STREET, NEW YORK, NY 10014.

Contents

Foreword

The Special Appeal of Charles Darwin

There have been many books about Charles Darwin in the century or so since his death, and it might seem presumptuous to offer another. But there is something special about both the man and his work that makes both of them eternally fascinating.

After all, how many of the original publications describing a revolution in science can be recommended as a good read for non-scientists? Isaac Newton's epic, the *Principia*, hardly qualifies (even if you do read Latin); Albert Einstein's papers on general relativity may look beautiful to mathematicians, but leave the rest of us less than enthralled. And the writings of the quantum pioneers are not something you would take to while away a train journey. True, James Watson did write a highly enjoyable book about the discovery of DNA; but that was long after the event, and dealt more with personalities than the science. No, there is only one candidate – Charles Darwin's *On the Origin of Species by Means of Natural Selection*, first published in 1859 and still in print in a Penguin volume that, significantly, reprints the *first* edition of Darwin's great work.

Darwin had more than the genius to make major scientific discoveries; he was also a superb writer, in the great Victorian tradition. He loved literature (one of the highlights of his social life was a meeting with George Eliot), and he agonised over his own prose, constantly rewriting his books, even at the proof stage, and driving printers and publishers to distraction with his last-minute changes. And yet, for all the frantic revisions, the result is a clear voice, talking to the reader in straightforward terms, describing the trials and tribulations of the research that led Darwin to his revolutionary conclusions, as well as explaining those conclusions as clearly as anyone has ever been able to do. And the object of that research is, ultimately, ourselves – the origin of humankind, along with all the other species of life on Earth.

It would be a dull reader who wasn't gripped from the opening passage of the book. The introduction itself is bound to hook anybody reared, like us, on a diet of *Coral Island* and *Swallows and Amazons*. 'When on board H.M.S. *"Beagle"* as a naturalist,' writes Darwin, 'I was much struck with certain facts in the distribution of the inhabitants of South America, and in the geological relations of the present to the past inhabitants of that continent. These facts seemed to me to throw some light on the origin of species – that mystery of mysteries, as it has been called by one of our greatest philosophers.' By the second paragraph, we learn that there is some urgency about Darwin's work, and that he lives under the threat of death. 'My work is now nearly finished; but as it will take me two or three more years to complete it, and as my health is far from strong, I have been urged to publish this Abstract.'

The 'abstract' ran to nearly five hundred printed pages in its original form, carefully setting out the story of variation among domesticated species and in the wild, ranging from pigeon breeding in London to the birds of the Galápagos Islands, before coming on to the heart of the matter, the struggle for existence. 'In looking at Nature, it is most necessary to keep the foregoing considerations always in mind – never to forget that every single organic being around us may be said to be striving to the utmost to increase in numbers; that each lives by a struggle at some period of its life; that heavy destruction inevitably falls either on the young or old, during each generation or at recurrent intervals.' By the time you reach the discussion of Natural Selection (Darwin always used the capitals) itself, you are well prepared. As you read about the 'preservation of favourable variations and the rejection of injurious variations, [which] I call Natural Selection,' you feel like the character in C. P. Snow's *The Search* (1934), who 'saw a medley of haphazard facts fall into line and order . . . "But it's true," I said to myself. "It's very beautiful. And it's true."'

In spite of the beauty and clarity of Darwin's original version, there are still debates about the nature of evolution. It is, perhaps, no surprise to find that many of the critics of the very idea of evolution have not, in fact, read the *Origin*; it is more of a surprise to find that even among biologists who debate exactly how evolution works there are reprobates who have somehow overlooked the importance of checking out what Darwin actually said. But someone must be reading the book. According to *The Portable Darwin* (edited by Duncan Porter and Peter Graham, Penguin, 1993), there were six editions, 35 printings and translations into eleven languages in Darwin's lifetime; the book has been continuously in print since, and the reprints now exceed 400 while the translations number at least 29 languages.

But one word of caution. Many of those reprints and translations are from later editions of the work, especially the sixth, revised by Darwin in the light of criticisms provoked by the first edition. Much of that criticism later proved to be ill-founded, and Darwin's accessible prose did, for once, suffer from his attempts to buttress his arguments. It is the first edition which strikes the reader with such clarity and force that it provokes the 'but it's true' reaction, and it is, happily, the first edition which most closely matches the way biologists today think that evolution really does work.

How, though, did Darwin come to write such a book? Where did his scientific ideas come from, and what kind of a man was it who could come up with a revolutionary new way of thinking about the world and then sit on it for years, not publishing his conclusions until his hand was forced by the realisation that another naturalist was thinking along the same lines? And what else did he do in his long career as naturalist, geologist, explorer and biologist? We hope to show you both Darwin the scientist and Darwin the man in this book, the quiet revolutionary who changed both science and society, but wanted nothing more than a placid, stable home life in the Kentish countryside. No doubt *The Origin of Species* will still be in print long after our book is forgotten, but meanwhile we hope we may introduce a few new readers to the delights of Darwin's work.

Michael White
John Gribbin
October 1994

DARWINS OF THE MOUNT, SHREWSBURY

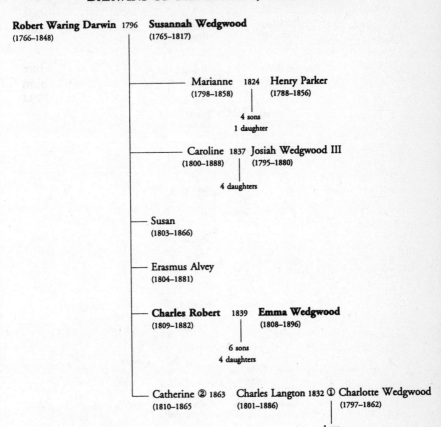

Robert Waring Darwin 1796 **Susannah Wedgwood**
(1766–1848)   (1765–1817)

Marianne   1824   Henry Parker
(1798–1858)   (1788–1856)

4 sons
1 daughter

Caroline   1837   Josiah Wedgwood III
(1800–1888)   (1795–1880)

4 daughters

Susan
(1803–1866)

Erasmus Alvey
(1804–1881)

Charles Robert   1839   **Emma Wedgwood**
(1809–1882)   (1808–1896)

6 sons
4 daughters

Catherine ② 1863   Charles Langton 1832 ① Charlotte Wedgwood
(1810–1865   (1801–1886)   (1797–1862)

1 son

CHARLES AND EMMA DARWIN'S GROWN UP CHILDREN

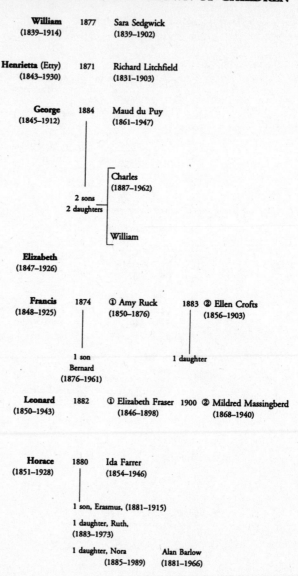

William 1877 Sara Sedgwick
(1839–1914) (1839–1902)

Henrietta (Etty) 1871 Richard Litchfield
(1843–1930) (1831–1903)

George 1884 Maud du Puy
(1845–1912) (1861–1947)

Charles
(1887–1962)

2 sons
2 daughters

William

Elizabeth
(1847–1926)

Francis 1874 ① Amy Ruck 1883 ② Ellen Crofts
(1848–1925) (1850–1876) (1856–1903)

1 son 1 daughter
Bernard
(1876–1961)

Leonard 1882 ① Elizabeth Fraser 1900 ② Mildred Massingberd
(1850–1943) (1846–1898) (1868–1940)

Horace 1880 Ida Farrer
(1851–1928) (1854–1946)

1 son, Erasmus, (1881–1915)

1 daughter, Ruth,
(1883–1973)

1 daughter, Nora Alan Barlow
(1885–1989) (1881–1966)

WEDGWOODS OF MAER HALL, STAFFORDSHIRE

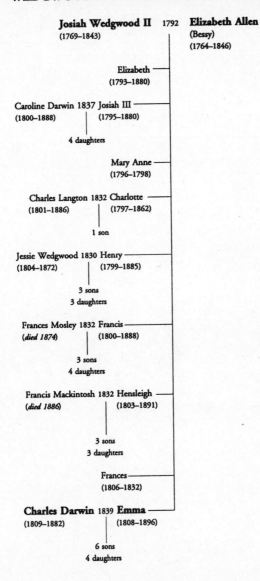

Josiah Wedgwood II 1792 **Elizabeth Allen**
(1769–1843)   (Bessy)
                              (1764–1846)

Elizabeth
(1793–1880)

Caroline Darwin 1837 Josiah III
(1800–1888)   (1795–1880)

4 daughters

Mary Anne
(1796–1798)

Charles Langton 1832 Charlotte
(1801–1886)   (1797–1862)

1 son

Jessie Wedgwood 1830 Henry
(1804–1872)   (1799–1885)

3 sons
3 daughters

Frances Mosley 1832 Francis
(*died 1874*)   (1800–1888)

3 sons
4 daughters

Francis Mackintosh 1832 Hensleigh
(*died 1886*)   (1803–1891)

3 sons
3 daughters

Frances
(1806–1832)

Charles Darwin 1839 **Emma**
(1809–1882)   (1808–1896)

6 sons
4 daughters

Chapter 1

Unsettled Youth

In a corner of Westminster Abbey dubbed Scientists' Corner lies a simple black slab of stone over which thousands of tourists daily walk. The stone bears the inscription:

CHARLES ROBERT DARWIN.
Born 12 February, 1809.
Died 19 April, 1882.

Beside Darwin's tomb lie those of William Herschel, Michael Faraday, James Clerk Maxwell and a few yards away stands a grandiose memorial to Sir Isaac Newton. Each day a tour guide explains how at the time of Darwin's death there were many who could not understand why he of all people should be buried in the most august church in Britain beside the kings, queens and great statesmen of the past thousand years. The abbey bookshop outside the church sells souvenirs and guides along with the great works of the many writers, philosophers and artists buried a few hundred yards away. *The Origin of Species* is conspicuous by its absence.

This odd ambivalence towards one of our greatest scientists is a thread which ran through Charles Darwin's life and has continued to this day. It was also an aspect of his own personality. Despite formulating one of the most influential and revolutionary works of science, Darwin himself could never proselytise his own ideas. Instead, he was championed by those who realised the importance of his work and were better suited to public speaking. Darwin was a great thinker but had very little self-confidence. He said in his autobiography:

> Whenever I have found out that I have blundered or that my work has been imperfect, and when I have been contemptuously criticised, . . . it has been my greatest comfort to say hundreds of times to myself 'I have worked as hard as I could, and no man can do more than this.'

Darwin was raised as a Christian, trained as a cleric and died an atheist. Yet, his theory of evolution does not dispute Christianity at its deepest level. Although it certainly destroyed fundamentalist principles and naive Creation myths, evolution is not an atheist doctrine as some would like to imagine.

Darwin was a very humble man, totally dedicated to his studies, a scientist who worked meticulously and in solitude. Benefiting from inherited wealth he never had seriously to pursue a career; but, despite his privileged upbringing he was a humanist before the term found popular appeal. Whig rather than Tory, Darwin argued for civil liberties and was a kind and attentive husband and father. Plagued by a succession of illnesses throughout the second half of his life, after returning from his famous expedition aboard the *Beagle*, he could work on his theories only in fits and starts and held back from publishing his greatest work for fear that it might be badly received.

When finally Darwin's most famous work, *The Origin of Species*, was published in November 1859 it resulted in further contradiction. To the many intellectuals who could not accept a belief in creationism or providential design, Darwin's revelations provided a clear, logical answer, but to most churchmen it came as a threat and a heresy. For popular consumption, certain sections of the press lampooned Darwin's work, instilling in the public consciousness a misleading idea of what Darwin and evolution were really about. These misconceptions are still very much with us today.

Withstanding the attacks of the Church and public misunderstanding, the theory of evolution has lasted and has long been seen as one of the basic tenets of biology. Along with the theory of relativity and quantum mechanics, it constitutes one of the fundamental principles which govern the universe. When Darwin conducted his pioneering work, the theory of evolution was ripe for discovery and had been postulated by many others, but this should, in no way, demean the man's achievement. In Darwin, all the factors necessary to develop the idea of evolution came together – a combination of experience aboard the *Beagle*, imagination, freedom to work and, perhaps most significantly, the influence of a strong family tradition of scientific interest and ability.

Charles Darwin was born into a post-revolutionary age. The process had begun in America in 1775, developed into the War of Independence of 1776 and crossed the Atlantic to inflame France in 1789. While Europe and America were engaged in violent reaction to the establishment, in Britain a more subtle revolution developed which would, in the long run, exert at least as much influence over the future of the world – this has since become known as the Industrial Revolution.

England sat at the epicentre of the Industrial Revolution, a process which most historians consider to have begun some time between 1760 and 1790. From England came a new age of steam power; iron and factories began to replace water power and wood, cottage industry and agriculture.

There was a new spirit abroad at the end of the eighteenth century. For a time, a new liberalism was fashionable, a new optimism which heralded the technological age, an early humanism, a faith in machines, science and philosophy. Coupled with this was an all-important sense of practicality. The Industrial Revolution gained its name because it altered everyday life. It was not a cultural change affecting only the rich, but it had its roots in the mundane.

Both of Charles Darwin's grandfathers fitted precisely the mould of the industrial revolutionary. His paternal grandfather, Erasmus Darwin, was a well-known doctor who had a country practice in Shrewsbury just north of Birmingham, close to what later became the industrial heartland of England. Erasmus was an intellectual who was as gifted in the field of literature as he

3

was in science. During the early 1790s and before Wordsworth, Blake or Byron had made their mark, Erasmus was considered to be the most famous English poet of his time. Coleridge called him: 'the first literary character in Europe and the most originally-minded man.'[1]

Erasmus was the archetypal gentleman polymath of his era. As well as being a dedicated physician, beloved by the scattered members of his huge rural practice, he dabbled in all areas of science and invention and has been credited with giving the idea of the steam engine to James Watt, a man he knew well. He also drew up designs for steam-powered vehicles and flying machines, designed and built canals and developed his own medical treatments.

Darwin's maternal grandfather was Josiah Wedgwood, the man who single-handedly, and starting with almost nothing, created the world-famous pottery business which until this day bears his name. He made a fortune supplying the highest quality tableware to royalty and nobility and sold one set to Catherine the Great of Russia for over £2,000.

The two men were close friends and active members of a group who called themselves the Lunar Society – a collection of wealthy men interested in machines and mechanical devices who met monthly at the time of the full moon. The Lunar Society flourished for a period between the 1760s and the early 1790s but when, towards the end of the eighteenth century, the social mood changed and turned against liberal thinking, it was disbanded, its most lasting legacy being the term 'lunatic'.

Both Josiah Wedgwood and Erasmus Darwin were political free-thinkers, who would today perhaps call themselves Social Democrats. They were also greatly influenced by Unitarianism, an unconventional form of Christianity which teaches that God has a single aspect (as opposed to the Holy Trinity). It is a doctrine which tries to eradicate the stifling dogmatism of legalistic Christian thinking. Such unorthodox social and religious views exerted a major influence on the two families for at least three generations and lay at the root of much of Charles Darwin's own attitudes to life.

Erasmus was married twice. His first wife, Mary Howard, died an alcoholic at the age of 30, by which time she had already born him five children. The fourth, Robert Waring Darwin, who was born in 1766, was Charles' father.

Josiah married one of his many Wedgwood cousins, Sarah, in 1764 and they lived happily together until Josiah's death in 1795. They had seven children, three daughters and four sons. The eldest child, Susannah, then married Robert Darwin in 1796.

At the time of Charles Darwin's birth in 1809, the Wedgwood and Darwin families had been closely linked for two generations. Erasmus and Josiah had met around 1760, even before Josiah had married, when both men were 30 and through their life-long friendship, members of the two families mixed socially. By 1796 Erasmus was an old man of 65 with only a few years left to live and Josiah had died a year earlier. Robert Darwin aged 30 had followed his father into medicine and had already made his mark. Charles' mother, Susannah, was a year older than Robert and had inherited £25,000 from her father. Robert and Susannah had known each other since childhood and were perfectly suited. Within the two families, their marriage was almost a foregone conclusion.

Robert Darwin had a large practice in Shrewsbury which had been set up for him by his father soon after he had qualified. At the time of his marriage to Susannah Wedgwood he bought a piece of land on the edge of Shrewsbury and built a large, rambling house he named The Mount. He lived there with his family until he died, aged 82, in 1848.

Robert Darwin was in many ways a larger-than-life character and had been forced into the role of his father's successor by a series of tragedies befalling the family. His mother died when he was four years old and his aunt Susannah, Erasmus' unmarried sister who helped to run the family home when Mary was alive, decided that she could no longer continue in this role. Robert was very attached to his aunt and was undoubtedly emotionally disturbed by the simultaneous loss of both her and his mother. Then, when Robert was twelve years old his elder brother, Charles, the apple of Erasmus' eye, died suddenly at the age of 20. Charles had been studying medicine at Edinburgh Medical School and had developed septicaemia from a cut finger obtained during a post-mortem of a child.

Erasmus' second son, Erasmus II, shared none of his father's interests and went into law. Consequently, although Robert was neither particularly academic nor, as a young man, overly interested in science, he bent to his father's will and followed a career in medicine.

As a middle-aged man Robert was seen to be in turns kindly and severe. He was very generous and accommodating with his family and friends, greatly respected and admired by both his colleagues and his patients, but, after his wife's death from peritonitis in 1817, he became morose and melancholy, his mood seeming to infect the atmosphere of The Mount. He never remarried.

Charles Darwin's early youth was a happy time for him. His parents were contented and wealthy and Charles, who was their fifth child, was pampered by his three elder sisters, Marianne, Caroline (who also married a Wedgwood, Josiah III) and Susan. Charles also had a brother, Erasmus (known by the family as Ras or Eras), four years his senior, with whom he was very close, and a younger sister, Catherine, born a year after Charles in 1810.

Until the age of eight Charles was educated at home by his sister Caroline, who was in her mid-teens and made the most of her role as strict disciplinarian. As a distraction from his sister's regime, Charles played solitary games in the vast family home. His father had become interested in the fashionable study of natural history and there were rooms full of exotic collections, stuffed animals and old bones. A massive greenhouse attached to the side of the house was a veritable jungle to a young boy and it was in this environment of learned eccentricity and an unforced seeking of knowledge that Darwin's fascination for natural history and biology began. Charles soon became a great hoarder, collecting anything that captured his interest, from shells to rocks, insects to birds' eggs. When he was not exploring the greenhouse or starting new collections, he fished for hours on the nearby River Severn.

In many respects, although he was contented with his hobbies, Charles' early childhood was a lonely time. The well-documented closeness between him and his brother Erasmus developed later. At the age of ten or eleven, a six- or seven-year-old brother is more of a liability than a bonus, and besides, Eras spent his term-times as a boarder at Shrewsbury School, leaving Charles the only boy among four sisters. To compensate, Charles began to seek attention. He made up tall stories, such as the time he insisted he had 'discovered' a secret cache of hidden fruits which he had in fact stolen from his father's treasured orchard and hidden in the grounds of The Mount.

In his own account of his childhood, written as an elderly man, Darwin is self-critical of his boyhood behaviour. With a sense almost of self-loathing, he recounts a time when he was cruel to a puppy. This incident haunted him his entire life because he could even recall the exact place, within sight of The Mount, where the incident occurred. As an adult, he was also honest enough with himself to confess that his reason for beating the puppy was the sense of power it gave him.

In reality, Darwin was probably much like any normal pre-school boy of his class. He sought attention and spent a lot of time existing in a semi-imagined world, but he was also cloistered from the harsher realities of the world. But then, after the death of his mother in July 1817, his whole life changed.

After Susannah's death, Charles' two eldest sisters, Marianne and Caroline, took over the running of the household. They were assisted by a staff of servants, who looked after the menial chores while the two girls supervised Charles and their younger sister Catherine. Charles' other older sister, Susan, who was almost fourteen, did her bit to help, while Erasmus remained at Shrewsbury School. Although it was undoubtedly an emotional shock to the family, because Susannah had been a virtual invalid for several months before her death, the two eldest girls were probably used to playing a major role in running The Mount whilst Dr Robert was out tending his patients.

It is clear that the death of his mother profoundly disturbed Charles and this was exacerbated by the fact that he was totally unable to express his sorrow and distress. With Susannah's death, Robert Darwin had fallen into a severe depression and pulled through the crisis only by throwing himself into his work. He often came home tired and in a foul mood and had decreed that no mention of Susannah's death should ever be made. This order it seems was passed down to Marianne and Caroline and they had imposed a wall of silence around the whole sad affair so that Charles was quite unable to express his emotions outwardly.

He could not turn to his father for comfort nor could he talk about his mother with his sisters.

According to modern theories of child psychology, this behaviour can be very damaging to young children and it has been speculated that this closing off of emotion permanently scarred young Charles. Several of Darwin's biographers have suggested that this inhibition lies at the root of the succession of long-term illnesses which plagued the adult Charles Darwin.[2] What is striking is that by Darwin's own account he can remember almost nothing of his mother and in particular the events surrounding her death. In his autobiography, he mentions her only briefly:

> My mother died in July 1817, when I was a little over eight years old, and it is odd that I can remember hardly anything about her except her deathbed, her black velvet gown, and her curiously constructed work-table.

He then goes on to describe his school life and childhood pranks.

A short time before his mother died, in the spring of 1817, Charles had become a day boy at a small school in Shrewsbury which was run by a Unitarian minister, Reverend Case, whose church Susannah had regularly attended for many years and in which the Darwin children had been baptised. He stayed there for little more than a year before being sent as a boarder to Shrewsbury School, a small but growing public school which later established itself as one of the most prestigious in Britain.

The headmaster at Shrewsbury School was the Reverend Samuel Butler, a strict disciplinarian who had transformed his school from a repository for the erring sons of rich gentry to an establishment already gaining a considerable reputation as a seat of learning, successfully preparing the sons of wealthy gentlemen for the great universities. It was strict and traditional at Shrewsbury and Charles hated it.

Charles detested regimented learning and had absolutely no interest in the Classics, which constituted the majority of the curriculum. He regularly cribbed off friends and did the absolute minimum to avoid a beating for laziness or poor test results. As Darwin himself put it:

> Nothing could have been worse for the development of my mind than Dr Butler's school, as it was strictly classical, nothing else being taught, except a little ancient geography and history. The school as a means of education to me was simply a blank.

The one great advantage about being at Shrewsbury was that his brother had been a boarder there for the past four years.

After the death of their mother, and with Charles' arrival at Shrewsbury, the two boys became very close. Neither of them enjoyed being there and Charles often spent evenings at The Mount, which was little more than a mile away across the fields. Dashing off after school and spending the evening in his own room and in the company of his family, however morose the atmosphere, seems to have been important to Charles. By all accounts Eras had long since settled into life at Shrewsbury School and had never been interested in making a habit of sneaking off home for the evening. Charles revelled in it and the race to get back to school before locking-up time took on the proportions of an adventure. He frequently dared himself to leave late but still make it back in time and, according to his own account, as he ran he used to pray to God for help in finding the speed to make it.

Although he still seemed attached to home, he got on well with the other boys at the school. Old friends recounted in later years that they always thought of the young Charles Darwin as old before his time and a very serious fellow, but he made friends easily and readily joined in with the social life of the school.

About the time he began at Shrewsbury, Charles took to going on long, solitary walks in the nearby countryside. During these treks he would go into a deep reverie thinking about all manner of things and often forgetting where he was. On one occasion he was so deep in thought that he fell into a ditch. It has been suggested that these solitary walks were another symptom of the heartache he felt over the loss of his mother and that it acted as a form of subconscious therapy.

Despite the fact that he grew into a sociable man, enjoyed many good friendships during his life and was a caring father and husband, Darwin always was something of a loner. It could be that his isolated nature developed from the early loss of his mother and the subsequent introspection it brought, or else the walks and the introversion could have been a natural part of his character. Darwin himself does not elaborate on this in correspondence or in his autobiography, so we can only speculate.

It was some time after Charles' tenth birthday that the first signs of an early fascination with science were shown. It was at about this time he went beyond simply collecting, hoarding what looked like pretty objects or playing games in the jungle of the greenhouse. In the summer of 1819 Marianne and Caroline took Charles to the Welsh coast for a holiday. Much to the annoyance of his two sisters Charles spent most of each morning wandering off on his own to watch birds or to hunt for insects. Hours later he would return with specimens and spend the rest of the afternoon and early evening bent over his finds, devising methods of cataloguing them and trying to ascertain the species to which the various creatures belonged. It was only after Caroline delivered a sermon to young Charles on the cruelty of killing creatures for his own interest that he gave up the practice and contented himself with collecting dead specimens.

Despite this restriction Charles continued to find a deep fascination in the ways of nature and to keep collecting and investigating. He was becoming increasingly gregarious at school and, finding that his teachers were largely uninterested in the things he found so fascinating, he actively sought out older boys who could help him learn more about nature. He also soon realised the potential of his father's huge library and spent many hours there poring over books about natural history.

It was soon after this that Charles and Eras teamed up in their scientific explorations. In 1822, when Charles was thirteen, Eras, now eighteen and in his last term at Shrewsbury, suddenly became obsessed with chemistry. Despite Charles' own interest in natural history, it was easy for Eras to convince his brother that chemistry was the noblest of all the sciences. Together they decided they would build a laboratory of their own in which they could emulate the great discoverers of the period whom they had read about and had discussed with their father.

It was a symbol of the boys' privileged upbringing that they could actually realise their ideas. They agreed to set up a fund for the laboratory and to pool their resources. Coercing their generous father with the idea that they were broadening their education, they very quickly managed to collect over £50, equivalent to a servant's annual salary in the 1820s. This was enough to build the lab and to equip it with the very latest in hardware, chemicals and furniture. The boys were quite aware of the fact that Dr Robert was a soft touch, particularly where an academic bent in his children was concerned, referring to him as 'the cow' to be 'milked'.[3]

Nothing appears to have come from the adventure except that it probably gave the boys some practical scientific experience which they would never have gained at school. A more immediate effect of the brothers' latest fad was a new nickname for Charles. For the rest of his time at Shrewsbury he was known simply as 'Gas'.

In the autumn of 1822 Eras left Shrewsbury to go to Cambridge. Charles now had the run of the lab and continued with his experiments alone. He spent most of his allowance on buying the latest gadgetry and chemicals for his solitary hobby and is even reported to have used sixpenny pieces as his source of silver, an extravagance over the top even for the son of a wealthy doctor.

Eras hated his work at Cambridge almost as much as his brother continued to detest the Shrewsbury curriculum. Eras found himself taught by dull lecturers and had to endure a seemingly endless succession of calcifyingly boring discourses. The brothers' correspondence was full of chemical chat and discussion of their shared hobby, leaving little room for comment on family matters and nothing but bitter references to their formal education.

At the end of Eras' first year at Cambridge, in the summer of 1823, Charles was allowed to visit him. It appears they had a rather raucous time and spent money like water. They almost certainly went out drinking and followed the rather gauche fashion of the day – inhaling laughing gas. Eras at nineteen was enjoying the life of a wealthy student sampling what he saw as the better things of life and following the fashionable trends of the day. He enjoyed attending poetry readings and fell in with an 'arty crowd'. Charles was easily influenced and looked up to his brother as a guiding light. Although Charles was probably a little young to be indulging in such adult pleasures, this hedonistic trip to Cambridge did no real damage and acted as a much-needed break from the Classics and the choking fumes of the lab.

Around this time a further break from Charles' academic preoccupations came in the form of a new interest – hunting. Although he greatly regretted this enthusiasm in later life, as a young man Charles displayed an insatiable desire to kill birds of any variety. At shooting parties organised by either the Darwins or the Wedgwoods, the fourteen-year-old Charles would rise early, employ a beater and be out in the field, with a dog, his gun loaded and ready before the rest of the household were even awake. Oddly, Marianne and Caroline remained silent on the morality of shooting birds. It was a peculiar obsession and one which the older, mature Charles Darwin could not fully understand except to suggest that his younger self was still power-crazed as he had been with the puppy many years earlier. This fervour for spending his time on sports rather than working, combined with a succession of very average school reports and a growing suspicion that Charles was not benefiting from the teaching methods employed there, led Robert Darwin to take his son away from Shrewsbury School. The summer term of 1825 was the last Charles was to spend there.

Charles had a deep love and respect for his father. Although Dr Robert himself had grown up in an emotionally impoverished environment, he was in adulthood a very considerate and caring man. After the death of his wife he became melancholy and moody but he always provided his children with everything they could wish for and was very concerned for their welfare. When he was able to shake off his own suppressed anguish over his loss, Dr Robert could be good company. Charles was particularly impressed by his father's wide-ranging interests and knowledge. Although he was no great scientist or medical theorist, Robert

Darwin was a well-read man with modern views and was, for his time, quite socially aware.

By the spring of 1825 Charles' attitude had become too much for his father. In a rare moment of anger Dr Robert told Charles that he 'cared for nothing but shooting, dogs and rat-catching' and that he would 'be a disgrace to himself and all his family.' That said, after removing him from Shrewsbury School, his father's next move was to have Charles spend the summer helping him with his practice. His intention was clearly to instil in the boy an interest in medicine. In this he was temporarily successful.

Charles took to medical practice with a natural flair and, astonishing as it may sound to modern sensibilities, by the end of the summer holidays the schoolboy was allowed to treat a small collection of his own patients, albeit under the watchful eye of his father. As well as developing his son's interest in medicine, Dr Robert was trying to bring out the compassionate side of Charles' character in order to awaken in him a sense that his former indulgences were not appropriate for a wealthy intellectual's son and not in keeping with the spirit of the family's liberal, humanist inclinations. It was proper for him to have a healthy interest in country pursuits – hunting and fishing – but not to the exclusion of almost everything else. In this the object lesson appears to have failed.

During the summer of 1825 Dr Robert was also busy planning Charles' academic future. Both Erasmus I and Robert Darwin had been students at Edinburgh Medical School and there had been a long tradition of links between the Darwin family and Edinburgh-educated intellectuals. Besides these factors, Edinburgh had a number of other advantages to offer. It was better equipped, employed superior tutors and, according to Darwin family opinion at least, it produced better-educated medics than the only other real option – Cambridge – and students who had been instructed in all the latest techniques imported from the continent mingled with the best traditions of British medicine. So, Edinburgh it was.

Luckily for Charles, Eras, who had now completed three years of his medical course, was about to take his external hospital study year, which, it was decided, would be in Edinburgh.

Doubtless, their father thought Eras would have a beneficial influence on his younger brother.

Although they played off each other's unconventional attitudes and disapproved of many of the tutors at Edinburgh, the two Darwin brothers were, in one sense, good for one another. They were both natural intellectuals and spent an inordinate amount of time reading the latest scientific, medical and political literature. When Charles should have been at lectures and Eras at hospital tutorials and demonstrations, they were usually to be found in their rooms or in one of the multitude of taverns beside a log fire reading extracurricular texts.

Edinburgh was a magnificent playground for a pair of wealthy young men in the 1820s. It was relatively cheap to live there and for them money was no object. They rented comfortable lodgings with the famous Mrs Mackay, who had seen generations of students pass through her rooms at 11 Lothian Street. They made a number of close friends, attended fringe lectures on all the latest theories and fads from the continent, and Charles frequently went off into the country or to the coast to collect specimens. He also attended private classes given by a freed South American slave called John Edmonstone, a talented taxidermist who lived a few doors away in Lothian Street. He worked at the Edinburgh Museum but gave private lessons to interested students in his spare time, charging only a guinea for a course. Charles attended Edmonstone's classes for an hour each day for two months and learned an invaluable skill for his later voyage aboard the *Beagle*.

During that first year in Edinburgh, from October 1825 until the summer of 1826, Charles appears to have had the time of his life. He continued to write home to his sisters and his father, but never breathed a word of what he and Eras were really up to. It is clear that Marianne and Caroline were suspicious of events north of the border because they often wrote to Charles reminding him of his responsibilities and piously referring him to passages from the Bible.

Charles and Eras somehow managed to scrape through the year at Edinburgh without drawing too much attention to themselves from the university authorities and delivered the minimum work necessary. Attracted by the unconventional, both brothers attended the lectures of one Robert Knox who was the black sheep of the Edinburgh tutor fraternity. Knox was considered to be a heretic by many of the more conservative teachers at the university because of his outspoken attacks against the orthodox church. However, many Edinburgh students, including the two Unitarian boys from Shrewsbury, were inspired by him. Unfortunately for Knox and perhaps another generation of potential students, he was disgraced some time during 1828 when he was found to have unwittingly accepted corpses from Burke and Hare. From then on his witty outbursts were silenced.

During his first year as a medical student, Charles learned a truth that would ultimately destroy any hopes he or his father might have had for a future career in medicine. After witnessing a particularly gruesome operation without anaesthetic conducted on a child, Charles realised that he was terminally squeamish. He hated dissection from his first day in theatre and was often physically sick when compelled to cut open a cadaver. The sight of a screaming infant being cut open before his eyes was too much. He fled the room and never again returned to the operating theatre. Worse still, the image of the dismembered child haunted him for the rest of his life. From that point on he consciously drifted through his time at Edinburgh doing everything but working on his official studies.

The summer holiday came as a very welcome relief for both boys. It was all very well playing the Bohemian intellectual but even they were beginning to feel guilty about having squandered their father's money and having done nothing in terms of official study. They were glad to return to the comforts of home and their family friends. Charles almost immediately slid back into his hobby of slaughtering small animals. When he was not out in the field shooting, he continued where he had left off, gathering specimens for his collections. By now he was travelling further afield and went on another trip to the Welsh coast. When the weather was bad, he and Eras would occasionally return to the lab they had built four years earlier or studied together in the library. During the evenings, when the atmosphere at The Mount became too much, the young Darwins, including the four girls, would spend the warm summer evenings with the younger members of the Wedgwood family at their country estate of Maer or else with another set of family friends, the Owens at their country home, Woodhouse.

In the autumn Charles had to return to what he by now considered to be his totally redundant studies in Edinburgh, while Eras set off for London where he was enrolled at the Great Windmill Street Anatomy School to begin the next stage of his medical training.

Alone in Edinburgh, Charles once more threw himself into a round of intellectual pursuits far removed from his official studies. Continuing to deliver the bare minimum of work required, he joined a group called the Plinian Society whose members met regularly to read each other papers on natural history. The Plinians saw themselves as rather adventurous non-conformists who were interested in the purest forms of academia. In reality they were not so unorthodox as they liked to imagine. When one of the members, a rebellious, 21-year-old student called William Browne delivered a talk based on his own version of materialistic philosophy, including a virulent attack on the Creation, his speech was struck off the society's minutes.

With time on his hands, Charles continued to pursue his interest in practical natural history. With others from the Plinian Society he visited the Firth of Forth and joined the trawlers dredging the ocean floor. His particular interests at the time were sea sponges, corals and sea-pens. Also through the influence of the Plinians he began what was to become a life-long fascination with geology.

Geology was still a relatively young science in the 1820s and almost nothing was known of the mechanisms through which the earth had evolved. The subject was taught from two opposing camps at Edinburgh University. The radicals of the Plinian Society preferred the teachings of Robert Jameson, who was regarded as a 'Neptunian geologist' as he believed that rock strata had been precipitated from a universal ocean. His rival was the more orthodox and conservative Professor Thomas Hope, who theorised that rocks had cooled from molten material. In contrast to his friends, Darwin preferred Hope, but conscientiously attended lectures by both men. In so doing he gained a greater insight into the subject which, combined with lengthy exploratory trips with his Plinian colleagues, meant that during his two years in Edinburgh, Darwin learned far more about geology than he did about medicine. As we will see, this provided excellent preparation for his serious studies in the subject at Cambridge under a master of the subject, Adam Sedgwick.

Soon after arriving back in Edinburgh to begin his second year studying medicine, he made a close friend of Robert Edmond Grant, who was also a member of the Plinian Society. Grant was sixteen years older than Charles and had practised as a doctor for a number of years before leaving the profession to study marine life. He was a rather sad character with a bitter streak which had made him extremely cynical. Many found his personality too difficult to cope with, but for some reason, Grant and Darwin immediately hit it off.

Grant was extremely knowledgable, widely read and had travelled all over Europe. He was a good influence on Darwin's intellectual development. A radical thinker, he opened Charles' eyes to some of the more extreme views of the day. From Erasmus I, the Darwins had been a family absorbed by religious ideology and, spurred on by the ideas of his new friend, Charles became fascinated with the subject.

Grant was concerned with what he viewed as patent contradictions between the contents of the Bible and the new findings of science. He introduced Charles to the risqué notions emanating from the continent and developed by intellectuals both in Edinburgh and at the newly founded University of London, known in some quarters as the 'godless college'.

Robert Grant's was precisely the kind of rebellious soul Darwin found so inspiring. In retrospect it is easy to see how he had a significant influence on Darwin's later thinking, particularly during the period when he was developing the theory of evolution. Grant did not subscribe to the generally accepted view of the day, according to which the fossil record came about by a series of divine interventions which allowed for intermittent evolution of nature. He had no idea what had happened but by even suggesting that the orthodox view was at fault it laid the ground for Darwin when he approached the subject years later.

Despite the fact that Darwin had enormous respect for Grant and learned a great deal from him during his second year at Edinburgh Medical School, they later drifted apart, largely because Grant found it almost impossible to compromise his views in the face of others' arguments. He was dogmatic and difficult to deal with socially; many of the intellectuals with whom he came into contact had no time for the man. As a result, his ideas, many of which were revolutionary and decades ahead of their time (see Chapter 2), were simply ignored.

Darwin's second spell in Edinburgh turned into a fiery year full of political theorising and self-motivated research. By the early summer of 1827 this impetus had run out of steam. The Plinian Society had fallen into disarray, Grant moved to London to accept an academic post at the University of London, Browne moved to France to pursue his revolutionary doctrines and Darwin finally decided that he could no longer continue with the charade of student life in Edinburgh and left without a degree.

In May, nervous at the prospect of returning home to face the music, Charles took himself off on a short trip around Scotland and then travelled straight on to London. It was his first visit to the capital and he was given a guided tour of the city by his cousin Harry Wedgwood, who was a newly qualified barrister. He then decided to cross the Channel to visit Paris, where it was arranged that he should meet up with Jos Wedgwood, Josiah I's son, who was joining his daughters, Fanny and Emma, just returned from the South. Before meeting the Wedgwoods, Darwin teamed up with William Browne and another friend from the Plinians, John Coldstream. The three of them tarried in the French capital for a few days, where the wealthy and carefree Darwin footed the bill for a succession of expensive meals, drinks and entertainment. Later, having eventually seen the Wedgwoods, Darwin noted in a letter home how Emma was even more beautiful than he had remembered from parties at Maer. In truth it may have been the first time Charles had really taken any serious notice of his cousin, and it set the tone for their friendship and eventual courtship many years later.

Finally mustering enough courage to face his father, Charles returned to The Mount in August. By then both father and son had had plenty of time to think about Charles' future, but it appears that, unlike Charles, Dr Robert had come to some conclusions over the matter.

Having evidently failed in his attempts to push Charles into medicine, Dr Robert had decreed that his errant son should study for the Church. He was aware that Charles had little time for orthodox religion, but he was now at his wits' end. Unless something was done Charles seemed destined for a life drifting through meritless hobbies or squandering the family fortune on his friends' expensive habits. What Charles needed, Dr Robert had decided, was a little discipline. Furthermore, with a degree from Cambridge under his belt, Robert could set Charles up in a comfortable position as a country clergyman. At the time the Anglican Church regularly sold off desirable jobs to the highest bidder. With such a position came the respectability Dr Robert wanted for his son. It would also provide Charles with a suitable cover for what was perceived as his incurable laziness and eccentric, often contradictory obsessions with natural history and hunting. To this end he insisted that Charles apply to Cambridge University.

Charles asked for some time to think about his options. This his father granted him, but there was no escape from Dr Robert's demands. Charles had to toe the line and to follow instructions. So the summer of 1827 was spent reading standard religious texts, many of which Charles found extremely hard to swallow, and trying to relearn the rusty Latin and Greek which he had dodged at Shrewsbury School. There was some relief from the hard work. Once again the Wedgwoods threw parties at Maer and the Owens invited him to Woodhouse. Charles spent his free time joining the hunt or going shooting. In the evenings he flirted with Emma Wedgwood at Maer or else visited Fanny Owen at Woodhouse. About this time he and Fanny began to see a lot more of one another and soon developed a serious liaison. Charles frequently rode over to Woodhouse between sessions trying to improve his Latin and Greek for the forthcoming entrance examination.

Whether it was the distraction of Fanny or simply lack of aptitude, it soon became evident that Charles' knowledge of the Classics was so poor that Dr Robert had to employ a private tutor for him during the late summer. In the autumn of 1827 Charles finally took the entrance examination and just scraped through. A few months later, soon after New Year 1828, he was on the move again, this time to begin a degree at the University of Cambridge.

For many the Cambridge of the late 1820s was a rather oppressive place, for others it was a veritable playground of pleasure. Unlike Edinburgh, Cambridge was a city totally dominated by the university authorities who not only had draconian powers over its 2,000 students but could also control the livelihood and social activities of the townsfolk. The university had two representatives in the House of Commons, while the town did not have a single spokesman; and the Chancellor, Vice-Chancellor and his 'police force', the Proctors, had powers to bar students from mixing with tradesmen (as a result of which the townfolk's businesses could be ruined or, if the offence merited it, the townsfolk could be imprisoned).

When Darwin arrived in Cambridge a term late, the town had just recovered from a series of riots where townsfolk and students conspired in an effort to change the rules of the university by force, attacking the Proctors' headquarters or conducting rowdy demonstrations beneath the Chancellor's windows.

Charles had little time for the intrigues of Cambridge politics and had been forewarned by his brother that if he stuck to a set of essential rules he would keep out of trouble. These were to obey the college curfew, never to be seen drunk in public, never to become involved in fisticuffs or to be seen out with a woman and never, ever to be caught out of gown.

Charles began life quietly at one of the smaller colleges, Christ's, and took lodgings above a tobacconist's in Sidney Street. For his first year he was an obedient and unassuming student. He made a few acquaintances and only one really close friend, his second cousin William Darwin Fox, who was also studying for the Church and was in his final year. Fox shared Charles' interest in natural history and had a number of collections which he was eager to show Charles. Fox was a member of a rather tepid group of students who were neither brilliant and triumphant in exams through total dedication nor were they lazy pleasure-seekers. According to inherited wisdom, Cambridge University consisted of only two varieties of student, 'the reading man' who pored over his texts and passed his degree with honours and 'varmint' who were usually from noble families, had money to burn, frequented gambling houses and taverns and were often caught in compromising positions with local whores, eventually scraping through their degrees by working slavishly in the final term of their third year. Fox's set, with whom Darwin became chums, fitted neither category.

Charles' first year passed uneventfully. He found himself back in Shrewsbury for the summer feeling that he had hardly experienced college life and as bored as he had been in Edinburgh. There was only one thing that saved his summer and indeed his next eighteen months at Cambridge – a new obsessive fascination with entomology, and in particular, beetles.

During the late 1820s a beetle craze was sweeping the country. Books about entomology were selling well and collectors vied with each other over who had the most comprehensive collections. Serious students of the subject spent days, sometimes weeks, on expeditions to various parts of the country in order to catch rare and unusual species which they killed and pinned to their displays kept in expensive cabinets and specially made cases. It was a hobby for wealthy gentlemen and was fired with a competitive atmosphere not unlike that now found in train-spotter fraternities or philatelist clubs. Not only was it a hobby which Darwin pursued with uncharacteristic competitiveness and vigour throughout his college years and beyond, but it also prepared him for his later work on board the *Beagle*. On occasion the new hobby could even be dangerous. On one trip when he was beetle hunting with a group of friends near Cambridge, Charles spotted two rare species of beetle on a tree. Clamping his palms over each, he was about to net his prize when he saw a third, even rarer variety on the ground. Momentarily at a loss as to what to do, without thinking, he popped one of the beetles from the tree into his mouth and with his free hand went to snatch up the third. Only then did he realise that the insect in his mouth was a bombardier beetle which squirted a repulsive-smelling, burning liquid into his mouth. Darwin panicked, spat it out and dropped the other two beetles.

For a while, Darwin's only distraction from beetle hunting was Fanny. By all accounts she was a beautiful young girl and a bit of a tomboy. She loved riding and playing billiards and wrote Charles rather risqué letters. From his replies, it appears that for a time at least, Charles was totally besotted with her. Charles and Fanny had a fiery, on-off relationship for much of the time they were together. Perhaps the fact that they met so infrequently added to the excitement. Their favourite pastime was to ride off into the woods of Squire Owen's estate, hunt for beetles and end up in a passionate embrace under the elms. From the comments she made in her letters to Charles when he returned to college, it was quite evident that Fanny enjoyed these illicit afternoons in the woods as much as Charles. In these letters she refers to her sadness at his absence declaring that she had 'nobody to ride with' and that if he did not come to visit soon (and perhaps referring to their games of billiards), she would soon forget her 'fine strokes'.

Back in Cambridge, starting three weeks late after the summer break, Charles was finding it almost impossible to settle in. He was still over-enthusiastic about shooting having rekindled his interest over the summer. Straight after arriving at college, he treated himself to a double-barrelled shotgun costing £20. When he should have been attending classes, he visited his brother Eras in London and on his return to Cambridge he continued to miss lecture after lecture.

During the autumn of 1828 Charles fell ill for a time with what appears to have been a serious infection of the lip. The source of this infection remains a mystery, but speculations about it range from a disease caught during his beetle trips while grovelling in piles of rotting vegetation to herpes. Wherever it came from, the infection flared up sporadically and continued to plague him throughout his college years. It become particularly troublesome during times of stress, which, because of his lackadaisical attitude, came often.

Towards the end of his second year Charles was beginning to get into serious trouble and even received an official warning from his tutors that he was heading for failure. It was only during his third year that he managed to pull the irons from the fire to save his degree at the last minute.

Charles' career was rescued by a number of coincidental factors. First, some time into his third year he split up with Fanny. It appears that, not to forget her 'fine strokes', she had been seeing another man while Charles was in Cambridge. For his part, during the autumn and winter of 1829 Charles had shown little real interest in her. By then he was far too busy trying desperately to make up for lost time even to write. He had repeatedly ignored her invitations to visit Woodhouse during the Christmas holiday and instead visited London or met friends in Cambridge between bouts of concentrated study in his rooms. The end came in the summer of 1830, soon after completing his third year, when Charles discovered that Fanny had become engaged to a wealthy clergyman called John Hill.

Charles was not overly disturbed by the loss of Fanny. By the third year he had at last begun to realise that this was his final chance and that he could not let his father down again. He was also beginning to take seriously the question of his future and although he had little time for orthodox Christianity and even less for Anglicanism, he could see that the Church could offer him a comfortable life-style and a suitable front to pursue his unabated interest in natural history and entomology. At the same time he fell under the influence of a number of people whom he greatly admired who could help him towards a successful completion of his degree. Chief among these new friends was a Cambridge don, Proctor, botanist and brilliant lecturer, the Reverend John Stevens Henslow, who became Charles Darwin's closest Cambridge friend.

Henslow was a brilliant man. He had become Professor of Botany at Cambridge by the age of 26, had then taken Holy Orders and accepted the curacy of Little St Mary's Church. This provided him with a substantial salary to supplement the modest annual £100 he was paid as a university professor.

It was Henslow who first really opened Darwin's eyes to the magic of botany and the 21-year-old soon became the middle-aged professor's favourite student. Henslow perfectly filled the gap in Darwin's life which up to then had been only partially satisfied by his obsessions with shooting, beetle hunting and Fanny. Now at last he had found an inspirational figure who turned him on to subjects that fascinated him and who would keep him within the curriculum of his official studies.

Darwin's last Christmas at Cambridge was spent cramming for the finals set for the third week of January 1831. It was bitterly cold, but Charles was hardly aware of anything going on around him. Only during the past year had he done much work at all and now his earlier laziness was returning to haunt him. Leaving his rooms only to visit Henslow in the Botanical Gardens, where the professor lectured in more clement weather, they wandered the gardens wrapped in winter coats, scarves and hats and then settled by the fire in Henslow's house. Henslow helped to guide Darwin through the torturous volumes he had to understand and answer questions on in the finals. These included *Principles of Moral and Political Philosophy* and *Evidences of Christianity*, both by the Reverend William Paley, who had died a quarter of century earlier and who followed an orthodox Anglican doctrine Darwin found hard going. Aside from this he had to complete questions on mathematics, a subject with which he had always struggled, as well as papers on philosophy, history and physics.

The exams were spread over three days, with papers each morning and afternoon. Charles had never worked so hard in his life and by the end of it he was exhausted. Thankfully he did not have to wait long to see how well or how badly he had done; within a few days the results were published. To his astonishment his hard work in the final year had paid off. He was placed tenth in the pass list of 178 students.

Darwin had finally graduated from Cambridge and was now well and truly on the road to joining the clergy. Little did he know as he walked away from the notice board announcing the results that he would never come closer to the Church than he was at that moment. Still less did he know that before the year was out he would be a member of an expedition that would change his life for ever and with it alter the course of modern biology.

Notes

Unless otherwise specified, quotations in this chapter are from *The Autobiography of Charles Darwin and Selected Letters*, 3 vols, ed. Francis Darwin, John Murray, London, 1887.

1 Desmond King-Hele, *Erasmus Darwin: Scientific Source for the Romantic Poets*. Talk presented to British Association for the Advancement of Science, 1993.
2 John Bowlby, *Charles Darwin: A New Biography*, Hutchinson, London, and Norton, New York, 1990.
3 *The Correspondence of Charles Darwin*, 8 vols, ed. F. Burkhardt and S. Smith, Cambridge University Press, 1985–93, Vol. 1, pp 1–5.

Notes

Unless otherwise specified, quotations in this chapter are from *The Autobiography of Charles Darwin and Selected Letters*, 3 vols, ed. Francis Darwin, John Murray, London, 1887.

1 Desmond King-Hele, *Erasmus Darwin: Scientific Source for the Romantic Poets*. Talk presented to British Association for the Advancement of Science, 1993.

2 John Bowlby, *Charles Darwin: A New Biography*, Hutchinson, London, and Norton, New York, 1990.

3 *The Correspondence of Charles Darwin*, 8 vols, ed. F. Burkhardt and S. Smith, Cambridge University Press, 1985–93, Vol. 1, pp 1–5.

Chapter 2

Evolution before Darwin

Charles Darwin was not the first person to think about evolution. He was not the even first member of his own family to think about evolution. But he was the first person to come up with a theory of evolution that proved able to withstand rigorous scientific testing. For a century and a half Darwin's theory of evolution by natural selection (modified slightly as our understanding of genetics has developed) has been the best theory of evolution we have had, and it is likely to remain so for the foreseeable future. In that sense modern biology began with Darwin's work; but it is still worth looking at some of the 'prehistory' of biology, to set Darwin's achievement in its proper, dramatic context.

It is a surprise to many people to learn that ideas attempting to describe the origins of the forms of life we know on Earth today in terms of evolution were around some two and a half thousand years ago, at the time of the Ancient Greeks. They were hardly mainstream views, and although the idea of evolution surfaced time and again in the centuries that followed, it was never to become established wisdom until the nineteenth century. Nevertheless, from at least the fifth century BC onwards there were people, from time to time, who tried to describe the origins of life in evolutionary terms.

One of the first of these thinkers whose ideas have survived to reach us was Empedocles of Agrigentum, who lived from 495 to 435 BC. The nuts and bolts of his ideas concerning the origins of living things are downright bizarre; but he did hit upon one very important element in evolution.

Empedocles actually suggested that plant life arose first after the formation of the Earth, and that animal life 'budded off' from the plants. But the plants did not produce whole animals. Instead, according to Empedocles, *parts* of animals budded off from the first plants – heads, arms, eyes and all the rest, as individual components, which then got together to create all kinds of weird and wonderful creatures. There were, on this picture, animals with the heads of men, two headed monsters, centaurs and all kinds of strange creatures. But, said Empedocles, the monstrous forms could not find mates and reproduce, and so they soon died out – became extinct, as we would say today – leaving behind only those creatures in which the various components worked together in harmony.

Here, amidst a welter of strange ideas linked to the myths and legends of the time, we find the first hint of the notion that evolution involves natural selection, with only the fittest forms (that is, the ones capable of reproduction) surviving. As Aristotle, who lived from 384 to 322 BC, commented in his *Physics*, Empedocles was the first person to suggest that the fittest forms of life (in the modern Darwinian use of the term) could have arisen through chance, rather than by design.

Aristotle himself came up with the idea of a chain, or ladder, of evolution, leading from the simplest sea creatures to fish, then to land animals, and on 'upwards' to humankind. He believed that this was an example of a natural progression in which Nature herself sought to achieve perfection – starting with the most imperfect of creatures, moving through a series of imperceptible transitions to the most perfect form of life on Earth, ourselves. 'Nature does nothing without an aim,' said Aristotle. 'She is always striving after the most beautiful that is possible.' And it seemed natural to philosophers at the time to regard humanity as the ultimate perfection, the most beautiful form that Nature could achieve.

This purposeful striving for perfection is, of course, quite different from Empedocles' idea of chance variations and survival of the fittest. Indeed, Aristotle set out Empedocles' theory (with his customary clarity and precision) only in order to refute it! So when Aristotle tells us that according to Empedocles:

> It is argued that where all things happened as if they were made for some purpose, being aptly united by chance, these were preserved, but such as were not aptly made, these were lost and still perish,

putting the idea almost in the language of modern science, he is doing so only in order to dismiss it and argue instead that there is 'a purpose in things which are produced by, and exist from, Nature'. The irony is that it is actually a combination of Empedocles' ideas about chance and survival of the fittest, together with Aristotle's idea of gradual evolution from simple forms of life to more complex forms (though *not* up a ladder of increasing perfection) that could have formed, more than three centuries before the birth of Christ, a workable version of what we know as Darwin's theory of evolution.

But Empedocles still had his supporters, including Epicurus, who was nineteen when Aristotle died, and who picked up the idea that many strange forms of life could arise directly from the earth, but that only those capable of reproduction have left descendants to be alive today. The Roman poet Lucretius, who lived in the first century BC, also promoted the idea, although he made no contributions of his own to the theory (if it can be justified by the name in this crude form) and even took a step backwards from Aristotle (like Epicurus) by rejecting the idea of gradual changes and describing more or less bizarre forms of life being created directly from the earth by spontaneous generation.

Once again, there was a better explanation already available at the time. Back in the fifth century BC, Anaxagoras, who lived from 500 to 428 BC (and was therefore a contemporary of Empedocles), had suggested that plants and animals formed from pre-existing 'germs' in the air. Like most of his contemporaries, Anaxagoras thought that the world had been designed and created by an intelligent being (or beings). But he left plenty of scope for chance and free will to play a part in what happened after the Creation, suggesting that these germs of life were brought to fruition in the warmth of the primordial slime, and their progeny then left to get on with life in their own way.

Yet another key element in the modern story of the origin and evolution of life seems to have been tantalisingly within the grasp of the Ancients. If only, the modern reader is tempted to say, Lucretius or one of his contemporaries had put this idea together with Aristotle's chain of gradual evolution and Empedocles' survival of the fittest – but remember that we have the advantage of hindsight, and that these ideas were just some of many possibilities discussed in the centuries before Christ.

In the centuries immediately following the birth of Christ, however, something remarkably like a modern view of evolution, but drawing on Aristotle's ideas and the notion of Nature striving for perfection, was accepted and taught by some of the founding fathers of the Christian Church. Gregory of Nyssa (331–396) taught that the Creation was potential – that God imparted to matter its fundamental laws and properties, but that the objects and completed forms of the Universe then developed gradually, under their own steam, out of primordial chaos. Saint Augustine (353–430) painted an even clearer picture. He taught that the original germs of living things came in two forms, one placed by the Creator in animals and plants, and a second variety scattered throughout the environment, destined to become active only under the right conditions. He said that the Biblical account of the Creation should not be read as literally occupying six days, but six units of time, while the passage 'In the beginning God created the heaven and the earth' should be interpreted:

As if this were the seed of the heaven and the earth, although as yet all the matter of heaven and of earth was in confusion; but because it was certain that from this the heaven and the earth would be, therefore the material itself is called by that name.

Augustine likens the Creation to the growth of a tree from its seed, which has the potential to become a tree, but does so only through a long, slow process, in accordance with the environment in which it finds itself. God created the *potential* for the heavens and earth, and for life, but the details worked themselves out in accordance with the laws laid down by God, on this picture. It wasn't necessary for God to create each individual species (let alone each individual living thing) in the process called Special Creation. Instead, the Creator provided the seeds of the Universe and of life, and let them develop in their own time.

In all but name, except for introducing the hand of God to start off the Universe, Augustine's theory was a theory of evolution, and one which stands up well alongside modern theories of the evolution of the Universe and the evolution of life on Earth.[1] His views were influential throughout the Middle Ages, and followed by such important thinkers as William of Occam (in the fourteenth century) and, most importantly, by Saint Thomas Aquinas in the thirteenth century. Aquinas simply quoted Augustine's teaching on the subject of the Creation and the interpretation of *Genesis*; but as he was one of the highest authorities in the Christian Church at the time, and has been one of the most influential since, this amounted to an official seal of approval for the idea that God had set the Universe in motion and then rested.

In a full thousand years, however, from the time of Augustine to the time of William of Occam, there had been virtually no progress in Christian Europe in thinking about the origins of living things – just as there had been virtually no progress of any kind in Europe during those Dark Ages. It was in the Arabic world that science and philosophy made progress at that time, and one side-effect of this was that for a time the Christian Church rejected the Augustinian view of evolution, for no better reason than that it drew on Aristotle's teachings, and Aristotle was held in high regard in the Muslim world, perceived then as a bitter opponent of Christendom.

Aristotle was translated into Arabic early in the ninth century, and his works had an enormous impact on the flowering of science in the Muslim world in the centuries that followed. To give just one example of how far ahead the Arabs were of their Christian contemporaries at that time, here is a passage from the philosopher Avicenna (980–1037), on the origin of mountains:

> Mountains may be due to two causes. Either they are effects of upheavals of the crust of the earth, such as might occur during a violent earthquake, or they are the effect of water, which, cutting for itself a new route, has denuded the valleys, the strata being of different kinds, some soft, some hard. The winds and waters disintegrate the one, but leave the other intact . . . It would require a long period of time for all such changes to be accomplished . . . but that water has been the main cause of these effects, is proved by the existence of [fossil] remains of aquatic and other animals in many mountains.

Avicenna did not realise that the other mechanism he describes, upheaval of mountains involving earthquakes, could also explain how the remains of fish come to be found high in the mountains today – because the rocks from which those mountains formed used to lie at the bottom of the sea. But, writing a thousand years ago (and some nine centuries before Darwin), he highlights two of the key elements that, as we shall see, were to set Darwin himself thinking along the lines that were to lead to his theory of evolution. The first is an echo of Aristotle's ideas about the evolution of life – that *gradual* processes, of the kind we can see going on today, are all that you need to explain how the world got to be the way it is. The second, which goes hand in hand with the first, is that a very long time indeed is required to explain all the changes that have occurred on our planet to make it what it is today.

When scientific thinking in Europe began to make progress once again in the sixteenth century, many of the ideas from which it developed came from the Arab world. In many cases, the original Greek texts of the ancients survived only because they had been copied into Arabic, and were later copied into Latin and then other languages. So it is no surprise to find Giordano Bruno, who lived from 1548 to 1600 (and ended up burnt at the stake for heresy because of his support for the idea that the Earth moves round the Sun) imbibing a mixture of

Arabic and Greek ideas, arguing for gradual changes in Nature rather than sudden cataclysms, and being led to the conclusion that the Earth must be much older than the few thousand years implied by a literal interpretation of the chronology in the Bible.

But the official line at the beginning of the seventeenth century is more accurately represented by the writings of Francisco Suarez (1548–1617), a Spanish Jesuit whose religious and geographical background made him particularly hostile to Muslim teachings. Suarez emerges as one of the founders of the idea of Special Creation, denying the possibility of any form of evolution from one kind of living thing into another and repudiating Saint Augustine's teaching on this subject and Saint Thomas Aquinas' acquiescence in Augustine's views.

Special Creation – the notion that each form of life on Earth was individually created by God, and has remained unchanged since its creation – became the accepted wisdom of the Christian Church, and was the standard teaching of the Church from the middle of the sixteenth century to the middle of the nineteenth century. With the establishment of Darwin's theory of evolution, the Church returned to the Augustinian interpretation of the Creation that had held sway for the previous thirteen hundred years. In truth, the idea of Special Creation, even within the context of Church teaching, dominated for less than three centuries out of fifteen between the time of Augustine and the time of Darwin (and out of two thousand years of Christianity to date). Even then, it was not entirely unopposed. But through the teaching of Suarez and his contemporaries (and their immediate successors), fuelled by an innate hostility to ideas that had been filtered through Muslim philosophers, the Christian Church was temporarily set in a mould of rigid opposition to the idea of evolution exactly during the time when scientific evidence in support of the idea began to emerge in Europe.

The conflict did not become apparent, though, until the middle of the nineteenth century. Before then, as scientific learning revived in Europe in the sixteenth and seventeenth centuries, the background to evolution was sketched in by a new kind of natural historian, in many cases (especially in England) men who were also ordained ministers of the Church. They had the time to study nature, and the inclination to understand better God's handiwork. A belief in Special Creation was no barrier to these attempts to understand the relationships between different forms of life on Earth today.

But in the hands of at least some of these naturalists, the attempt to understand life on Earth became increasingly scientific, in the modern sense of the term. Following Galileo's discoveries in physics and astronomy early in the seventeenth century (with their practical implications for navigation), the new breed of natural philosopher began to appreciate the importance of checking their ideas about the natural world by careful observations and experiments. It was no longer acceptable to suggest, for example, that different pieces of animal bodies could be produced by plants, and could then get together to form all kinds of bizarre animal; now, theories had to be based on observations of what really went on in the world.

One of the early champions of the new philosophy was Francis Bacon, who lived in England from 1561 to 1626. His approach was typical of the spirit of the investigators of the time, and he held that both Nature and the Bible were the work of God, so that the study of Nature (God's work) was as important as the study of the Bible (God's word) in understanding God. In his *Advancement of Learning*, published in 1605, Bacon wrote:

> To conclude, therefore, let no man out of a weak conceit of sobriety, or an ill-applied moderation, think or maintain, that a man can search too far or be too well studied in the book of God's word, or in the book of God's works; divinity or philosophy; but rather let men endeavour an endless progress or proficiency in both.

The quote is particularly noteworthy since Charles Darwin, anticipating an adverse reaction to his theory of evolution from some members of the Church, used it opposite the title page of the first edition of the *Origin of Species*.[2]

Bacon planned to produce an encyclopedia of all knowledge, but this was never completed. Some of his work on this did appear in 1620, as *Novum Organum*, and in this he provides some of the earliest speculations about variations of species. He talks of 'errors of Nature', in which 'Nature deviates and turns from her ordinary course' to produce minor variations or monstrous offspring, and points out the importance of understanding the normal form of a species in order to detect and study such variations. And he then goes on to spell out the use people make of the natural variability of species in breeding animals and plants for their own purposes – artificial selection:

It would be very difficult to generate new species, but less so to vary known species, and thus produce many rare and unusual results. The passage from the miracles of Nature to those of Art is easy; for if Nature be once seized in her variations and the cause be manifest, it will be easy to lead her by Art to such variation as she was first led to by chance; and not only that, but others, since deviations on the one side lead and open the way to others in every direction.

Writing in 1620, more than two hundred years before Darwin set sail on the *Beagle*, and more than three hundred years before the discovery of DNA, Bacon provided as neat a description of the potential of genetic engineering as you could wish to see. But even without the *cause* of natural variations being manifest, he was spelling out the way farmers take advantage of the chance variations among individuals that occur in nature and use them selectively to breed strains of wheat that yield more grain, or strains of pig that produce more meat, or strains of cow that give more milk.

It was not just in England that this revolution in thinking about the world was taking place. In France, René Descartes (1596–1650) made the daring suggestion that everything in the Universe could be explained in terms of 'a few intelligible and simple principles upon which the stars, and earth, and all the visible world might have been produced' – although, admittedly, he did hedge his bets, no doubt for the benefit of the Inquisition, by adding '(although we well know that it has not been produced in this fashion)'. In Germany, Gottfried Wilhelm Leibniz (1646–1716), best remembered as a mathematician and one of the inventors of calculus, was also interested in the living world. He thought about the relationship between fossil ammonites and the nautilus, which lives in the ocean today and looks like a smaller version of an ammonite. Ammonites, he wrote:

> are reckoned a kind of Nautilus, although they are said to differ always both in form and size, sometimes indeed being found a foot in diameter, from all those animal natures which the sea exhibits. Yet who has thoroughly searched those hidden recesses or subterranean depths? And how many animals hitherto unknown to us has a new world to offer? *Indeed it is credible that by means of such great changes [of habitat] even the species of animals are often changed.**

Many of the seventeenth and early eighteenth century natural philosophers considered similar views to these. The idea that the Universe is governed by a set of intelligible laws; the idea of variation among individual members of a species; even the possibility of some form of 'survival of the fittest'. But the key to further progress at that time lay in Bacon's insight into the importance of studying living species and understanding them, in order to be able to understand the variations.

The first significant attempt to produce a systematic classification of plants and animals – a taxonomy – was made by the English naturalist John Ray, who lived from 1627 to 1705. He went up to Cambridge in 1644, and was a student during the turmoil of the English Civil War. After graduating, he became a Fellow of Trinity College, trained as a clergyman and was eventually ordained in 1660. But he lost his position at Cambridge in 1662, following the restoration of the monarchy, when he refused to sign an oath promising conformity to the established doctrine of the Church. As far as natural history was concerned, this proved a blessing in disguise.

After losing his job, Ray was supported by an affluent friend, Francis Willughby, who shared his interest in natural history. They toured Europe together, studying the flora and fauna and

* Our emphasis. The 'new world' to which Leibniz refers is, of course, America.

collecting specimens, and on their return to England began to collaborate in publishing an account of what they had found. Their plan was that Ray would categorise and describe the plants, while Willughby would be responsible for the animals. In 1672, however, Willughby died at the age of 37, and the whole task fell upon Ray's shoulders. For several years, he stayed as tutor to Willughby's children, supported by a legacy from Willughby and marrying the governess of the children. In 1678, however, he and his wife returned to Ray's childhood home in Essex, where he lived for the rest of his life.

Ray published an enormous amount of material, most of it under his own name but some with Willughby listed as author or co-author, although he cannot have made much of a contribution before his untimely death. In *Ornithology*, first published in 1676, they described birds, and grouped them together, in terms of a classification based on their structural features and habits – land and water birds, birds with curved beaks and birds with straight beaks, fruit eaters and insect eaters, and so on. They identified individual species and gave them names, both English and (usually) Latin. In his *Natural History of Plants*, a three-volume epic which appeared between 1686 and 1704, Ray came up with a definition of just what a species is. 'After a long and considerable investigation,' he wrote in the first volume, 'no surer criterion for determining species has occurred to me than the distinguishing features that perpetuate themselves in propagation from seed.'[3]

In other words, pea plants produce seeds which give rise to other pea plants, oak trees produce seeds which give rise to other oak trees, and although there may be differences between the offspring and their parents, these differences are never enough to justify the offspring being called a different species. There is natural variation within the individual members of a species. Ray was also an early supporter of the idea that fossils are the petrified remains of dead animals and plants, a notion that was not generally accepted until nearly a hundred years later.

Ray's classification covered more than 18,000 species of plant, describing their appearance, where they lived, and general features of the life cycles of the plants. It paved the way for the work of Carolus Linnaeus, probably the most famous of all taxonomists, in the eighteenth century.

Linnaeus was a Swedish botanist, also known as Carl von Linné, who lived from 1707 to 1778 and used the latinised version of his name (in much the same way that he established the use of Latin names for plant and animal species). In the short time since Ray's ground-breaking work, there had been an explosive growth in information about living things, and Linnaeus had twice as many known species of mammal, for example, to categorise. He did so using a technique which formalised the use of two Latin names, and which as well as doing the invaluable task of placing individual plant and animal species in their appropriate family groups clarified the way the biological world works.

In the *System of Nature*, first published in 1735, Linnaeus spelled out the point aired by Ray, that offspring resemble their parents, or like begets like. It seems odd to modern eyes that this needed spelling out at all, but just over two and a half centuries ago there was still a widespread view that almost any living thing could give rise to almost any other living thing – deformities born to domesticated or wild animals, for example, were usually 'explained' in folklore of the time as a result of matings between members of different species, and the notion that some creatures appeared spontaneously out of warm earth was still prevalent. Linnaeus said that no new species are being produced today, and that an egg (or seed) always produces offspring closely resembling the parents.

He also realised that most parents produce more than two offspring, so that there is a tendency for populations to increase. Working backwards, this means that there were fewer individuals in preceding generations, and taking this to its seemingly logical conclusion he argued that each species on Earth today is descended from a single pair of ultimate ancestors. The Creation of those original representatives of each species he ascribed to God. This left no room in Linnaeus' eyes for evolution; he thought that species were unchanging once created. But just a century later, the importance of increasing populations became, as we shall see, a key ingredient in Charles Darwin's theory of evolution.

Linnaeus described plants and animals in a system involving a four-level hierarchy. The entire kingdom of plants was divided and subdivided into class, order, genus and species. Later, the category 'family' was added between order and genus,* but the basic classification is still used today. He formalised the use of Latin names in 1753, improving the more vague usage of Latin terminology by Ray and others in previous generations. In this 'binary nomenclature' the first Latin name in the pair is given a capital letter and specifies the genus, while the second (without a capital) specifies the species. The names are also usually put in italics. So the lion becomes *Panthera leo*, while the tiger is known as *Panthera tigris*; they are members of the same genus, but are different species.

The most dramatic feature of Linnaeus' classification was that he included human beings, as *Homo sapiens*. In his original classification there was a second member of the genus *Homo, Homo troglodytes*, but this was to prove a fiction based on tall stories about wild men. At the time, Linnaeus received some criticism for daring to include humans with the brute animals; he responded by challenging his critics to show him any feature of the human body which differed from that of other animals. Nobody met the challenge.

Today the criticism might be reversed – it may be that humans are not special enough to be placed in a genus all by ourselves. But in the middle of the eighteenth century the notion that humans might be classified like the animals, and studied like the animals, was a major step forward. What's more, Linnaeus did not try to arrange species in a ladder with humankind at the top; people really were classified as just one species among many.

Linnaeus also puzzled over the geographical distribution of species, and how they interacted with one another and the physical environment. Although accepting the idea of a Creation, he dismissed the idea of taking the story of Noah's Ark literally as ludicrous, and suggested instead that all living things might have originated from a single mountainous island surrounded by ocean. This ingenious idea tried to provide a suitable original

---

* For some specialist uses there are further subdivisions, but we do not have to worry about them for our story.

environment for each species by placing the mountain near the Equator, and making it tall enough so that at some altitude up the mountain conditions would be just right for each form of life on Earth (with polar bears at the top of the mountain). But it never overcame the difficulty of explaining how the members of different species could come down the mountain and travel to their present-day homes, passing through many unsuitable environments along the way (think of the difficulties faced by those polar bears!). The more people began to understand the ecological balance of living things, and how well each species fits in to its local environment, the more incredible this idea of migration from the site of the Creation became.

Late in his life, Linnaeus modified his ideas about the perfection of species slightly. The original species created by God might, he said, have been modified slightly in recent times, and the number of species increased, as they were 'multiplied by hybrid generation, that is, by intercrossing with other species'. But his tentative shift away from the idea that species were fixed and unchanging went nowhere near as far as the ideas of his exact contemporary the Frenchman Georges Louis Leclerc, Comte de Buffon (1707–88).

Buffon was one of the first popularisers of science. Educated in law, mathematics and astronomy, he inherited a fortune at the age of 25, and was able to spend his life following his scientific interests without having to worry about money. He became Keeper of the Royal Garden in Paris in 1739, and began writing what resulted in a 44-volume description of the natural world. Although Buffon did not make many original contributions to science, he was helped by several naturalists of his day to coordinate a wealth of information about the living world, and he wrote in a clear and interesting style. His *Natural History* was widely read by educated people across Europe.

Like Linnaeus, Buffon worried about the geographical distribution of species. Comparing the fauna of South America with the fauna of Africa, he pointed out that none of the animals in the torrid zone of one continent was found in the equivalent environment on the other continent, and concluded that instead of migrating from a single point of origin each species had arisen in the region where it is found today. He did not dismiss migration entirely, however, and suggested, for example, that the North American bison might be descended from a form of ox that had migrated to the New World from the Old World at a time when the world was warmer than it is today. 'Having afterwards advanced into the temperate regions of this New World,' he said, 'they received the impressions of the climate and in time became bisons.'[4]

This, clearly, implies some form of evolution, and includes the insight that species evolved to fit their environments. He also drew attention to the way in which many animals are built on body plans which include traces of rudimentary organs or limbs that are of no use to it – such as the vestigial legs possessed by some snakes. He was particularly severe on the pig, which, he said:

> Does not appear to have been formed upon an original, special, and perfect plan, since it is a compound of other animals; it has evidently useless parts, or rather parts of which it cannot make any use, toes all the bones of which are perfectly formed, and which, nevertheless, are of no service to it. Nature is far from subjecting herself to final causes in the formation of these creatures.

This line of argument can be extended to overturn Paley's famous 'argument from design'. That holds that the perfection of living things shows that they are the work of a Designer. In fact, though, many living things (like the pig) are a hodge-podge of bits and pieces, adapted essentially by trial and error to serve functions quite different from their original use. No intelligent designer could possibly take credit for this mess – and it is certainly hard to see how anyone who has witnessed the birth of a human baby can imagine that our own bodies were designed by some super intelligence.

Buffon himself, however, could not quite shake off the idea that species were originally produced by a Creator, and he talked of the 'degeneration' of the original forms into the forms we see today. The ass, for example, was seen as a degenerate form of horse (although the horse itself, running about on its fingernails, can hardly be said to have a logically designed body plan), the bison was seen as a degenerate form of oxen, and the apes were seen as degenerate forms of humankind.

But although some of Buffon's conclusions were vague and contradictory, groping towards an idea of evolution while still clinging to the notion of Creation, his writings helped to influence many naturalists to travel the world on long and dangerous voyages, collecting specimens and studying the distribution of species in different environments. The scene was being set for Charles Darwin's own voyage of discovery.

Buffon himself also thought about the history of life on Earth, and the history of the Earth itself, which was to be another key input into Charles Darwin's thinking – more of this in Chapter 4. But before Charles Darwin had even been born, his grandfather Erasmus, writing at the end of the eighteenth century, came up with his own theory of evolution, a theory which was developed independently a few years later by the Frenchman with whose name it is more commonly associated, Jean Baptiste Lamarck.

Erasmus Darwin, who lived from 1731 to 1802, appears as rather a shadowy figure in many biographies of his grandson, and we do not have space to do him full justice here – his life was varied and intriguing enough to merit a full-length biography of his own, at least as long as the present book. But we can at least highlight some of the key points in his life. As well as being a medical doctor of sufficient skill and fame to be asked by George III to become his personal physician (Darwin declined the post), and an active natural philosopher, elected a Fellow of the Royal Society in 1761, Erasmus Darwin was a poet of such stature and popularity in the late eighteenth century, as we saw in Chapter 1, that in 1790 there was a serious possibility of his becoming the Poet Laureate. His work is regarded as having had a significant influence on the English Romantics;[5] his literary circle also included Anna Seward, the 'Swan of Lichfield'.

Rather strangely to modern eyes, even some of his scientific ideas, including his thoughts about evolution, were sometimes expressed in the form of long poems. Even more strangely, one of these long poetical works, *The Botanic Garden* (itself influenced partly by the work of Linnaeus), has a distinctly erotic flavour. In the light of that, however, it may be less surprising to learn that as well as his conventional families (he married twice, producing five children by the first marriage and seven by the second, the last being born in 1790) Erasmus Darwin had a long-standing relationship with a 'Miss Parker', and was the father of her two illegitimate daughters, born in the early 1770s. They grew up to run a boarding school for girls, which led Erasmus to publish *A Plan for the Conduct of Female Education in Boarding Schools* in 1797.

Very much a free-thinker, Darwin and the Lunar Society (mentioned in Chapter 1) supported both the American and the French revolutions, and spoke out against slavery. This led to attacks from politicians such as George Canning (who was later briefly Prime Minister), contributing to a decline in Darwin's popularity as a poet.

Erasmus Darwin's ideas about evolution appeared in his two-volume prose work *Zoonomia*, published in 1794 and 1796, and in the verse work *The Temple of Nature*, which appeared posthumously in 1803. Although Darwin's evolutionary ideas were rather buried in *Zoonomia*, a large work mainly devoted to other matters, their publication in the middle of the 1790s is an important indication of just how far ahead of his contemporaries were his ideas. The more complete expression of those ideas can hardly be said to have been presented to the world with a flourish either, appearing posthumously and wrapped up in poetical form. But careful unravelling of the two works gives a clear picture of the way Darwin thought evolution worked.

Darwin was familiar with the ideas of Linnaeus, Buffon and others, but went beyond them to suggest how evolution might work. He accepted the old Greek idea of spontaneous appearance of life on Earth, but only in the simplest form, as 'specks of animated earth'. He then paints a picture of life moving from the sea on to the land, and evolving more complex forms as it does so:

> Organic life beneath the shoreless waves
> Was born and nurs'd in ocean's pearly caves;
> First, forms minute, unseen by spheric glass,
> Move on the mud, or pierce the watery mass;

These, as successive generations bloom,
New powers acquire and larger limbs assume;
Whence countless groups of vegetation spring,
And breathing realms of fin and feet and wing.

Even today, you could hardly improve on that as a description of the emergence of life – and the verse form certainly makes it more memorable than most modern text books! Darwin illustrated his ideas with reference to changes that occur during the development of living creatures (for example the change from a tadpole to a frog), and included the suggestion that humankind had evolved from monkey-like ancestors – all this before Charles Darwin had even been born. In later verses, Erasmus Darwin describes the fierce struggle for existence, with animals fighting one another and plants competing with one another for soil, moisture, air and light; but he just missed making the connection between this struggle and evolution that was to make his grandson famous. Instead, he suggested that characteristics acquired by individuals in their struggle for life were passed on to their descendants:

All animals undergo transformations which are in part produced by their own exertions, in response to pleasures and pains, and many of these acquired forms or propensities are transmitted to their posterity.

In other words, if a hungry bird has to probe deep into crevices with its beak in order to reach tasty insects, according to Erasmus Darwin's idea that bird would develop a slightly longer beak (in effect, by willpower), *and the descendants of that bird would inherit the slightly longer beak*, giving them a head start in the next generation. The striving of the offspring to probe deep into cracks will make their beaks infinitesimally longer still, and so on. This clear statement of the idea that evolution depends on the transmission of acquired characteristics appeared in the *Zoonomia*, and was the first published suggestion of this possibility.

The mechanism is now known to be wrong, and seems silly to many modern eyes. After all, to take the classic example, the children of a blacksmith are not born with extra large muscles.

But at least Erasmus was trying to explain evolution within the framework of science, and the rest of his description of the slow accumulation of changes was precisely two generations ahead of its time. Organs such as the rough tongue of a cow, so useful when eating grass, had, he said, 'been gradually produced during many generations, by the perpetual endeavour of the creatures to supply the want of food'. And in one of the most remarkable passages of all in the *Zoonomia*, he suggests that life began as a single 'filament':

> Shall we conjecture that one and the same kind of living filament is and has been the cause of all organic life? . . . I suppose this living filament, of whatever form it may be, whether sphere, cube, or cylinder, to be endowed with the capability of being excited into action by certain kinds of stimulus.

To modern eyes, this almost reads like a decription of DNA, the molecule of life. A rather accurate image of this would indeed be as a filament, a long thin molecule, capable of being excited into action by certain kinds of stimulus; and some biologists believe that all life on Earth may be descended from a single molecule of DNA that appeared out of the primordial soup. Erasmus Darwin was a *very* farsighted visionary.

His ideas about the mechanism of evolution were so similar to those of Lamarck that there is no need to go over them again in placing Lamarck in his historical context. Lamarck was born in 1744, thirteen years after Erasmus Darwin; before the French Revolution his name was Jean Baptiste Pierre Antoine de Monet, Chevalier de la Marck. As the name suggests, he came from an aristocratic family, but one with little money. After army service, he went into medicine and then became interested in meteorology and botany, being admitted to the Academy of Science in 1778. He became a protégé of Buffon and was appointed Botanist to the King, Louis XVI, in 1781 and later worked at the Royal Gardens. He survived the turmoil of revolution, becoming Professor of Zoology at the Museum of Natural History in Paris, with a suitable adjustment of his name.

In view of the situation in France and the wars with England, it is not surprising that Lamarck may have been unfamiliar with the *Zoonomia* when it first appeared, and it is probable that his own evolutionary ideas, which were first published in 1801, were arrived at independently. But there are striking similarities between his work and that of Erasmus Darwin – so much so that some people have suggested plagiarism. On the other hand, as we shall see, there was an equally striking parallel between the evolutionary ideas of Charles Darwin and Alfred Wallace half a century later, and in that case there is no doubt at all that the two men arrived at their conclusions independently. Coincidences do happen in science, especially when the time is ripe for new ideas. It is doubly odd, though, that the Darwin–Lamarck coincidence should so strikingly foreshadow the Darwin–Wallace coincidence.

Lamarck deserves more credit than he is often given for his ideas, which are sometimes presented as if they consisted only of the clearly ludicrous notion of the inheritability of acquired characteristics. Of course, he did suggest, like Erasmus Darwin, that characteristics acquired during the lifetime of an organism are passed on to its descendants. One example he used was of a giraffe, stretching its neck to reach succulent leaves and producing long-necked offspring as a result. But he did at least accept, and promote, the idea that evolution could occur, and he said that species were not fixed and unchanging. He even appreciated that the idea of a ladder of evolution would not work, and that a better description of the relationships between species would involve a branching pattern. Also, although the detailed mechanism for evolution that he put forward was wrong, like Erasmus Darwin he realised the importance of environmental factors, such as the availability of food, in shaping the course of evolution. He may have invented the term 'biology' (he was certainly the first person to use it widely); and he came up with a cracking good definition of a species:

A species is a collection of similar individuals which are perpetuated by generation in the same condition, as long as their environment has not changed sufficiently to bring about variation in their habits, their character, and their form.

Lamarck's ideas were taken up and championed by, among others, Geoffroy Saint-Hilaire (1772–1844), who worked with Lamarck for a long time in Paris. Saint-Hilaire developed his own variation on the evolutionary theme, suggesting that the driving force was the direct influence of the environment on organisms, which produces changes that are then inherited.

If these modifications lead to injurious effects, the animals which exhibit them perish and are replaced by others of a somewhat different form, a form changed so as to be adapted to the new environment.

This is nothing less than the idea of survival of the fittest, published by a French biologist in the 1820s. Unfortunately, some of Saint-Hilaire's other ideas were not so prescient.

Early in the nineteenth century, culminating in the publication of his *Anatomical Philosophy* in 1818, Saint-Hilaire had carried out an elegant study of the skeletons of many different species of vertebrate, and had shown that they are all built on the same plan. But he got carried away by his success, and went on to try to prove that this common body plan applies to *all* animals. Using ingenious arguments, he claimed that the different parts of an insect's body correspond to the various parts of the vertebrate skeleton. He then tried to make a case that the same correspondence applies between the body plan of a vertebrate and the structure of a mollusc.

This utter tosh enraged Georges Cuvier, an influential member of the French scientific establishment (of whom much more in Chapter 4). In 1830, a year after Lamarck had died, he launched a scathing attack on Saint-Hilaire in which the notion of a single body plan for all species was thoroughly blown out of the water. The comprehensiveness of the assault also embraced Lamarck's ideas about evolution, which Cuvier had never liked, and with Lamarck no longer there to defend himself, they were tarred with the same brush of disgrace. Cuvier urged naturalists henceforth to confine themselves solely to describing the positive facts about the natural world, without attempting to explain the features of the natural world in terms of half-baked theories. What might have been a natural progression in France from the ideas of Lamarck through the work of Saint-Hilaire to something like the theory that was actually developed by Charles Darwin received a setback from which it never recovered. For, while the echoes of the assault by Cuvier on Saint-Hilaire and Lamarck were still reverberating around the French Academy of Science, in the year following that assault Charles Darwin himself set sail around the world on HMS *Beagle*.

Notes

Unless otherwise specified, quotations in this chapter are from Henry Osborn, *From the Greeks to Darwin*, Macmillan, New York, 1894.

1 See, for example, John Gribbin, *In the Beginning*, Viking, London, and Little, Brown, New York, 1993.
2 See, for example, the reprint by Penguin, Harmondsworth, published in 1983.
3, 4 Quoted by David Young, *The Discovery of Evolution*, Cambridge University Press, 1992.
5 Donald Hassler, *Erasmus Darwin*, Twayne, New York, 1973.

Chapter 3

The Beagle

Darwin almost never made the voyage on HMS *Beagle*. When the invitation arrived in the form of a letter from one of his Cambridge tutors, George Peacock, writing on behalf of his friend Captain Francis Beaufort of the Admiralty in London, Darwin was on a geological trip in Wales. Arriving back late on the night of 29th August 1831, he found a large envelope awaiting him containing the letter from Peacock and another from his friend and mentor Henslow, insisting, although there was no need, that Darwin should definitely take up the offer of accompanying the young Captain Robert FitzRoy on a circumnavigation of the world.

Darwin was immediately interested; that was not a problem. The difficulty lay in persuading his father that it was a worthwhile project. There was no time to waste, the ship was due to leave within a month, so, despite being exhausted from his trip to Wales, Charles decided that he would have to talk it over with his father that evening. As expected, Robert dismissed the idea out of hand. According to Charles,[1] his father rejected the scheme on the following grounds:

1. Disreputable to my character as a Clergyman thereafter.
2. A wild scheme.
3. That they must have offered to many others before me, the place of Naturalist.

4. And from its not being accepted there must be some serious objection to the vessel or expedition.
5. That I should never settle down to a steady life hereafter.
6. That my accommodation would be most uncomfortable.
7. That you consider it as again changing my profession.
8. That it would be a useless undertaking.

Darwin was desperate to accept the invitation but could do nothing without his father's support. For a start, he would have to pay for his place on the *Beagle*. He would need spending money for the duration of the trip, as well as the funds to buy expensive equipment, research materials and books. It is easy to understand Robert's fears for his son and his exasperation over the way Charles always seemed to stumble into the most eccentric schemes and plans. As far as Robert was concerned, he had finally pinned his errant son down to a career in the Church, when suddenly along came this ridiculous offer for him to travel around the world.

Charles could do little to change his father's mind, but his berth on the *Beagle* was saved by the intervention of Robert's closest friend Jos Wedgwood, who, on hearing about the project, invited Charles over to Maer, where they together composed a letter to Robert, outlining the plus points of the voyage. Robert had the utmost respect for his friend's judgement and saw Jos as the most level-headed man he had ever known. The fact that Jos Wedgwood supported Charles' plans encouraged Dr Darwin to think again, and after a long discussion about the trip with Charles at The Mount, he finally gave his approval.

There was no time to celebrate. With less than four weeks to the proposed sailing date, he had a mass of things to deal with. Before he could do anything he had to confirm that he would be accepted on the trip and made an appointment to meet Captain FitzRoy.

Captain Robert FitzRoy was the grandson of the Duke of Grafton on his father's side and the first Marquis of Londonderry on his mother's – a direct descendant of Charles II. He was 26 at the time of the voyage, a gentleman, reputedly a fine sailor and a highly competent captain. Educated at Harrow before joining the Royal Naval College at Portsmouth, he shared with Darwin the fact that he had lost his mother at an early age; FitzRoy's mother, Lady Frances Stewart, had died when he was five.

Taking over the command of HMS *Beagle* in 1828, by the time he first met Darwin in 1831, he had already commanded the vessel on two highly successful voyages to South America. The planned third trip was to complete her contribution to the South American Survey, the charting of the continent of South America and its environs, a project financed by the Admiralty Department in London. A detailed survey of the region was essential for the safety of the merchant fleet as it spearheaded the development of trade in largely uncharted areas of the world. The burgeoning British Empire was spreading its economic and colonial tendrils to girdle the world and the survey ships travelled in the vanguard.

Darwin and FitzRoy hit it off immediately. They were both highly educated young men (at 22, Charles was the junior by four years), they had similar tastes in many things and were both from socially elite backgrounds. FitzRoy was deeply interested in science and spent hours discussing biology and geology with Darwin. He had read the first volume of Charles Lyell's famous *Principles of Geology*, published in 1830, which was a cornerstone of Darwin's own ideas in the subject, and was himself a keen amateur naturalist. The only area in which the two men strongly disagreed was in the realm of politics. FitzRoy was a dedicated Tory and had failed as a parliamentary candidate in the recent general election, while Darwin and his entire family were actively Whig. Knowing FitzRoy's politics in advance of their first meeting, Darwin determined to steer clear of the subject, at least until they had got to know each other better and he had been accepted on the trip. Later their political differences often boiled over and caused furious rows on the voyage.

Soon after Darwin and FitzRoy met, Charles wrote to his sister Susan describing the Captain in gushing terms: 'It is no use attempting to praise him as much as I feel inclined to do,' he said, '. . . for you would not believe me.'[2] FitzRoy clearly felt the same way about Darwin. In a letter to his superior, Captain Beaufort, he said: 'I have seen a good deal of Mr Darwin, today having had nearly two hours conversation in the morning and having since dined with him. I like what I see and hear of him, much, and I now request that you will apply for him to accompany me.'[3]

Darwin knew that his invitation to join the ship was primarily to keep FitzRoy company. The trip was scheduled to last three years and the Captain needed to have a companion who was intellectually stimulating and of approximately the same age and class. The *Beagle* had an appointed surgeon, Robert McCormick, but in the 1830s naval surgeons were considered of low social rank, precluding him from sharing the captain's table. Both McCormick and FitzRoy were keen and competent naturalists so Darwin's role as ship's naturalist was, at first, of secondary importance.

For two weeks Darwin and FitzRoy travelled around London buying supplies for the trip. Darwin, who had to pay £500 for his place on the *Beagle*, also bought books, a telescope costing £5 and a rifle for £50 as well as measuring apparatus and writing materials. Nothing was provided for him by the Admiralty.

On Sunday 11 September FitzRoy took Darwin to see HMS *Beagle* in Devonport. When Charles saw the ship for the first time, he was less than impressed – it was tiny. 90 feet long and only 24 feet wide amidships, the *Beagle* was a ten-gun brig and when it began the voyage it carried 74 men. Apart from the fact that it was an ageing vessel, already eleven years old in 1831, rotting and undergoing a complete overhaul in Devonport that summer, there were only two cabins, one for the captain, situated below decks and another, 10 feet by 11 which had been allocated to Darwin and two others, John Lort Stokes, the Assistant Surveyor, and Midshipman Philip King. As well as accommodating three men, Charles' cabin contained a large chart table, three chairs, all his equipment, many of his books and the mizzen mast passed straight through the middle of the room.

Darwin had known that shipboard conditions were going to be cramped, but he was still shocked by this first tour of the ship. Having decided to play down his anxieties over the ship both to his family and to FitzRoy, he told his sister in a letter shortly after seeing the *Beagle* for the first time: 'The vessel is a very small one; three masted, and carrying 10 guns: but everyone says it is the best sort for our work, and of its class it is an excellent vessel; new but well-tried, and half again the usual strength.'⁴

Despite the fact that FitzRoy had intended to leave Devonport in early October, it was to be another three months before the *Beagle* finally set sail for South America. First, delays were caused by the acquisition of special equipment as well as last-minute fittings and repairs, then bad weather forced them back to port after two abortive attempts to leave.

During the lull, Darwin visited London on a couple of occasions. The city, and indeed most of the country, was in turmoil. The recently elected government was trying to force through the Reform Bill – a radical plan to alter the social framework of the country, to change the electoral system and to create a more powerful middle-class. But, after its third reading, the Bill was rejected by the House of Lords. The country erupted, there were mass demonstrations, 70,000 people marched in London and several Tory dukes and bishops were attacked in the streets. Matters got so bad that the newly crowned king, William IV, was forced to suspend Parliament. Amidst the chaos, Darwin wandered around with his head in the clouds, his mind occupied solely with the voyage ahead and the kaleidoscope of exotic destinations awaiting him.

Finally, after Charles had spent the autumn kicking his heels and trying to keep news of the delays from his father, on a clear sunny morning, 27 December 1831, the crew lifted the *Beagle*'s anchor for the third time and the ship headed out into the open sea, bound for their first stop in the Canary Islands.

Within hours of leaving port, Darwin was seasick and took to his hammock, where he stayed as the *Beagle* ploughed its way through heavy seas in the Bay of Biscay. He was so violently ill that he could not leave his cabin until 6 January when the ship entered the calmer waters around the island of Tenerife.

Darwin had been looking forward to visiting Tenerife and had even planned a private trip there before the offer of a berth on the *Beagle* had come his way. His interest lay in the geological formation of the Canary Islands with its rugged volcanic landscape and close proximity to the African continent. It was a manageable distance to travel from England and offered a wealth of opportunity for the enthusiastic naturalist and geologist alike. Ironically, Darwin never made it there. On 6 January, as the ship approached the port entrance, it was met by the local authorities informing Captain FitzRoy that because of a cholera outbreak in England, if they wanted to enter the port, the crew would have to be quarantined for twelve days. FitzRoy would have none of it and immediately ordered the crew to turn the *Beagle* about. They sailed back into the open sea, leaving the beautiful vista of Santa Cruz and the scientific treasures of the Canary Islands behind. Ahead lay the expanse of the Atlantic and another month of sailing broken only by a brief stop-over at the Cape Verde Islands (described in Chapter 6), tiny rocks dotted in the ocean on the way to South America. Darwin felt crushed.

Although he continued to feel seasick, there were moments of light relief. Along with 31 other members of the ship's company, this trip marked Darwin's first crossing of the Equator, and he had to undergo the traditional seafarer's ritual to celebrate crossing the line. Dressed in full costume, FitzRoy adopted the role of Neptune and each of the initiates were brought to him before being covered from head to toe in paint and pitch and ducked in a water-filled sail. Darwin was lucky to be picked first and because of his privileged position, he was given relatively mild treatment. The whole ceremony ended up as a huge water fight which even involved the Captain. According to Darwin's diary of 17 February: 'Of course not one person, even the Captain, got clear of being wet through.'

The *Beagle* arrived on the east coast of South America, weighing anchor at All Saints Bay, Bahia (now Salvador) on 28 February 1832. For the next two and a half years the ship was to travel up and down the coast, over to the Falkland Islands and beyond the southernmost tip of the continent to Tierra del Fuego. As they entered All Saints Bay, Darwin caught his first glimpse of jungle, one of the great exotic vistas he had been looking forward to since before leaving England.

The most arduous phase of the journey had been the first leg, from Devonport to the Canaries. After the storms of the Bay of Biscay, the weather had improved and the Atlantic Ocean had remained calm throughout the crossing. He had even grown accustomed to the cramped conditions on board ship, commenting in his diary, three days before arriving in Bahia:

> Since leaving Tenerife the sea has been so calm that it is hard to believe it is the same element which tossed us about in the Bay of Biscay. This stillness is of great moment to the quantity of comfort which is attainable on board. Hitherto I have been surprised how enjoyable life is in this floating prison.

Once he had arrived in South America, Darwin's work really began. He was entranced by the luxuriant variety of life in the jungle. Here there was a lifetime of work to be done and only a short time in which to do it before the ship was due to leave on the next stage of the voyage.

FitzRoy's task in Bahia was to take soundings of All Saints Bay and to construct a chart of the harbour. This would take them until 18 March, giving Darwin less than three weeks to gather samples and to make his scientific analysis of the region. He struck out into the jungle and noted everything he could. His drawing ability was poor, so he relied upon written descriptions, and began collecting notebooks, the first of what would amount to a huge collection by the end of the voyage. These notebooks were to act as the background for *The Voyage of the Beagle*, written many years later.

As well as being his first experience of the tropics, the journey was also his first direct encounter with slavery, a practice Darwin loathed. This was not a view shared by FitzRoy, a high Tory. The two men had views representing the polar opposites of the time. On the one hand, Darwin, sometimes described as a mutaphiliac because of his ability to embrace change, and on the other, FitzRoy, the mutaphobe, a man who wished to maintain the status quo, a fundamentalist Christian and a believer in the natural superiority of whites over all other races. Living together in such close proximity on board the same tiny vessel, arguments were inevitable.

Although for the most part, Darwin and FitzRoy got along well during the voyage and Darwin admirably fulfilled his job of keeping the Captain company, they had many quarrels during the five years they were together. Not surprisingly perhaps, they never did see eye to eye over politics and their greatest bone of contention was over the practice and moral basis of slavery.

Soon after their arrival Darwin experienced the full blast of FitzRoy's anger when, over dinner, they had their first heated argument about the rights and wrongs of the slave trade. After too much wine and a bitter row, the Captain told Charles that he would no longer be required to stay aboard the *Beagle* and stormed out of the cabin.

The other officers reassured Charles that the Captain was not serious. After all, the crew called him 'Hot Coffee' because of his tendency to boil over easily. They were right; before morning the Captain had sobered up, calmed down and sent an apologetic note to Darwin asking his forgiveness and taking back his comments. All was forgiven.

As this incident suggests, FitzRoy was an unstable character and indeed there was known to be a history of mental illness within his family. His uncle, Viscount Castlereagh, had cut his own throat in a fit of depression ten years earlier, in 1822. FitzRoy was aware of the potential dangers of being at sea for long periods, and to compound his anxieties there was the fact that the previous commander of the *Beagle*, Pringle Stokes, had committed suicide during an earlier voyage. It was for these reasons that FitzRoy had insisted on having an intelligent, gentleman companion on board in the first place – to preserve his own sanity.

With the task of surveying All Saints Bay completed, the *Beagle* set sail for their next stop, Rio de Janeiro, further down the eastern coast of Brazil. As the ship made port three months into the voyage, their first mail arrived from England.

Darwin had been homesick for some time and news, even if it was months old, was very welcome, but at the same time it made him feel even more isolated. The gossip had all been played out long before and he felt like an outsider, privy only to distilled, dog-eared accounts. Despite the fact that the voyage was a great adventure, for a time, Darwin could not help but feel stranded and out of touch.

Rio marked the first occasion when Darwin was left for any length of time to travel inland on a scientific expedition while FitzRoy took the *Beagle* off to continue his own work, and it was over two months before he was reunited with the ship. During that time he studied the wildlife of the Brazilian jungle and was particularly fascinated with the beetles and other insects living on the jungle floor. Here he was to find a whole host of species never seen in Europe, beetles to put to shame the best collections in England. When the ship did return to collect him, it brought with it bad news. A fever had taken the lives of three of the crew and the ship's surgeon, Robert McCormick, tired of living in Darwin's shadow, had left and returned to England.

It was from this point on that Darwin was considered to be the 'official' naturalist for the voyage. It did not alter his passenger status but did provide him with a new nickname coined by FitzRoy – *philos*, the ship's philosopher.

After spending the hot summer in the steamy jungle, Darwin was glad to be back on the *Beagle* and to feel the fresh air of the open sea. Although the Brazilian jungle had provided a feast of opportunity and a vast array of specimens, he disliked the wealthy Brazilians and European colonialists who owned vast estates and abused the hordes of slaves they used for manual labour. He had witnessed estate owners beating their slaves and splitting up families, selling them separately for profit. Darwin wrote in his diary: 'How weak are the arguments of those maintaining that slavery is a tolerable evil!' Darwin's personal experiences with slavery gave substance to what, within his family, had always been accepted wisdom. In the conclusion to the 1845 edition of *The Voyage of the Beagle*, he makes his feelings very clear. 'It makes one's blood boil, yet heart tremble to think that we Englishmen and our American descendants, with their boastful cry of liberty, have been and are so guilty.'[5]

As much as he hated it, he was powerless to do anything about the things he witnessed. He was not a social reformer, but a scientist. It was only through the medium of his later work that he could hope to educate. But as a 23-year-old at the outset of what was to be a life-changing voyage, he had no ideas about transforming the way people perceived the world. All that lay far beyond his imagining, in the distant future.

Heading south, the *Beagle* sailed into winter and as they passed through a severe lightning storm on the way to Montevideo, the temperature dropped dramatically into the low fifties. Darwin was captivated by the excitement of it all, reporting in his journal:

> We witnessed a splendid scene of natural fireworks: the masthead and the yard-arm-ends shone with St Elmo's light, and the form of the vane could almost be traced, as if it had been rubbed with phosphorus. The sea was so highly luminous that the tracks of the penguins were marked by a fiery wake, and the darkness of the sky was momentarily illuminated by the most vivid lightning.[6]

They reached Montevideo just as trouble broke out in the city. An insurrection involving a group of troops looked as though it might threaten Europeans living there, and FitzRoy was asked to play the role of gunboat commander. Fifty crewmen, armed and bursting for real action, quashed the uprising with little effort and the city was returned to relative calm. Darwin was with them on shore, but although he carried a couple of pistols tucked into his belt, he had been ordered by FitzRoy to keep well away from the action.

A few weeks later they stopped over at Buenos Aires and by the third week of August the *Beagle* had reached Bahia Blanca. This was real frontier territory, land owned by the Spaniards who treated the locals even more appallingly than the British did further north. The area was primed and ready to boil over politically, a situation aggravated by the greed of European colonisers who treated the indigenous population contemptibly. The colonisers were able to keep order only by superior military force, which meant that most areas outside the guarded settlements were unsafe for visitors. While ashore, Darwin and the others had to carry arms at all times and were advised not to travel alone.

They stayed in the Bahia Blanca area for two months. While FitzRoy continued with his own duties, Darwin took himself off inland to study the local flora and fauna and to research the geology of the region. In his spare time Charles managed to find time to shoot ostrich and armadillo and a huge indigenous rodent called an agouti.

The work here was exciting and quite unexpectedly provided him with his first discovery of fossil remains. The first find, some ten miles away from the ship, in a bay called Punta Alta, consisted of a thigh bone and some teeth from a huge mammalian creature. Stumbling upon the fossils during one of his solo ventures, Darwin removed them from the cliff face in which they were buried and carried them back to the ship. To his surprise and annoyance, FitzRoy was apparently far from impressed and failed to realise their scientific significance.

For his own part, although he knew they were important, Darwin had little idea from which creature the fossilised bones had come, suggesting in his notes that they belonged to a large animal similar to the only other fossil of this type he had heard of – a sloth fossil which had been acquired by the College of Surgeons shortly before he had left home. Returning to the site the next day Darwin found other bones from the same creature and gradually pieced together an image of a mammal about the size of a cow.

Before leaving aboard the *Beagle*, Darwin had arranged for his former tutor at Cambridge, John Stevens Henslow, to act as receiving officer for his finds and throughout the voyage he regularly sent home boxes of specimens by ship. Unsure of their value, Charles nonetheless cleaned the fossils, packaged them carefully and arranged for them to be transported back to England along with the other samples in the next delivery.

Communication was by no means all one way. At major ports all along the coast, the crew received mail and, albeit months late, they also managed to catch up with the latest political and social news from home, both via letter and by talking to local European colonisers.

Family letters informed Charles that his former sweetheart, Fanny Owen, had finally married (her engagement to John Hill had been called off just as Charles was preparing to leave England). By all accounts, the man she had married, an aspiring politician called Biddulph (who, perhaps out of jealousy, Darwin later described as a varmint-like character) was turning out to be an unwise choice. The news saddened him deeply. Despite the fact that he was seeing and experiencing things beyond the wildest dreams of his friends and family back home, he still felt as though life really was slipping away and that he was missing out on the natural flow of life in England.

Eras was continuing to lead a life of leisure and spent his days socialising, reading and dabbling in his own experiments, all of which appeared to have little motivation or direction. Within the realm of politics, events which had been reaching a climax before the *Beagle* had left port had been resolved. Thanks to a group of Tories switching sides and voting with the Whigs, the Reform Bill had finally been passed. Many thought that this had saved a revolution in England, but as the *Beagle* headed further south no one on board, neither Whig nor Tory, could say for sure what the long-term outcome of the move would be.

On 19 October 1832 the *Beagle* headed back north to Montevideo to take on supplies and mail. They then stopped at Buenos Aires, perhaps the most civilised and European of the cities of the east coast of South America, where Darwin shopped for cigars and scissors, notebooks and pens, attended church and visited a dentist. It was a lull in the storm of rugged landscapes and perilous exploration, the last stop-over in a modern port before setting sail for the cold southerly regions of this vast continent and the savagery of the distant outpost of Tierra del Fuego.

Captain FitzRoy's official task in the South Atlantic was to carry out the surveys as ordered by the Admiralty; but he also had a private, self-financed project to complete. During his last trip to South America commanding the *Beagle* the previous summer, he had visited Tierra del Fuego and returned to England with three natives of the islands. They had ended up on board after FitzRoy's crew had become involved in a skirmish on one of the islands. Apparently when they had first reached Tierra del Fuego, the shore party had been attacked by a band of savages who made off with their boat. In retaliation, FitzRoy had taken the families of the culprits hostage and gone after the thieves. A fight had ensued, one of the Fuegians had been killed (his body was later skeletonised and used as a medical aid upon their return to England), and after the release of the families, three of the Fuegians were kept aboard the *Beagle* because they could not be put ashore on their home island without further trouble.

Returning to England, FitzRoy had decided to use the Fuegians as part of an experiment to see if it was possible to 'civilise' primitive races. The three natives, two males and a female, had been given the names York Minster, Jemmy Button and Fuegia Basket. The eldest, York Minster, was thought to be in his late twenties, Jemmy about fifteen and the girl, Fuegia, was ten or eleven. On this return trip, FitzRoy intended to put his 'civilised savages' back into their home environment in the hope that they, with the help of an English missionary, Richard Matthews, who had volunteered for the task, would teach their countrymen European ways.

They reached Tierra del Fuego on 17 December 1832. Although during the journey south FitzRoy had described to Darwin the primitive people who lived there, he was still surprised by the savagery of the natives and the bleakness of the place. Arriving in the Bay of Good Success, the *Beagle* was saluted by a band of yelling, hollering natives half-hidden in the dense forest which trailed down to the water's edge. In his journal of that evening Darwin recounted his first meeting with the natives:

When we came within hail, one of the four natives who were present advanced to receive us, and began to shout most vehemently, wishing to direct us where to land. When we were on the shore, the party looked rather alarmed, but continued talking and making gestures with great rapidity. It was without exception the most curious and interesting spectacle I had ever beheld. I could not have believed how wide was the difference, between savage and civilised man. It is greater than that between a wild and domesticated animal, in as much as in man there is a greater power of improvement.[7]

FitzRoy planned to return the three Fuegians along with the missionary Matthews after the official work in the area had been completed and before leaving for their next stop, the Falkland Islands.

Christmas and New Year of 1833 were spent touring the cluster of bleak, wind-swept islands. Darwin continued to make his detailed observations of flora and fauna and when he was not collecting wildlife, he doggedly hammered away at rock faces. At one of the most southerly points, Hermit Island, the crew shot game from the deck to replenish their food supplies. Despite partaking in his favourite hobby, this period in Tierra del Fuego was for Darwin the most miserable of the entire trip. The ship was at sea for a large proportion of the time, travelling between islands, and he was again violently seasick. The deck was awash and storms raged around the islands, clustered as they were around one of the world's most deadly stretches of sea – Cape Horn – a seafarers' graveyard.

By the end of January the Admiralty work was complete and they dropped anchor at Woollya Cove, Ponsonby Sound, home of Jemmy. A small party including Darwin rowed to the beach with the three Fuegians and the missionary. Once ashore, the group started to build a small settlement, consisting of a group of huts. They then transferred stores from the boats to the buildings to provide Matthews and the returning Fuegians with supplies to get through the first few weeks as well as trinkets with which to pacify the natives.

On their first night, and even before the *Beagle* had left, the camp was attacked. No one was hurt and Matthews was determined to carry on. They were well armed and Matthews at least was certain that they were in no real danger, cheerfully waving goodbye to Darwin and the others as they rowed back to the ship.

The *Beagle* set off to complete one final sweep of the area. Nine days later they were back at the Cove to check on the progress of the missionaries. The camp was still in one piece – just; but they were met by a badly beaten and demoralised Matthews. The camp had been repeatedly attacked and eventually overpowered; their possessions had been stolen and the party was exhausted, scared and hungry. Understandably, Matthews had had enough. Giving up the project, he returned to the *Beagle* totally disillusioned.

With his original plan in tatters, FitzRoy decided to leave the three Fuegians to their own devices with the promise that the *Beagle* would drop by to see how they had fared during a planned return trip the following year. According to his journal, Darwin was melancholy at the thought of leaving the Fuegians with their countrymen. It appears that York Minster, who had married Fuegia Basket, was quite happy to set up home there with his young wife, and held no fear of his countrymen, but Jemmy did not share his optimism.

On the other hand, Darwin was delighted to be leaving the islands. The Falklands may not have presented the most appealing port of call, but it is clear from his journals that Tierra del Fuego was one of the most miserable places Darwin had ever experienced. In his accounts of the islands and their inhabitants he creates the impression that the Fuegians seemed destined to remain for ever in a primitive state, unable to drag themselves out of their savagery and to adopt what Darwin, FitzRoy and most educated people of the day believed to be the only true way to live – the civilised manners, behaviour and morals of the European. Although coming from opposite ends of the political spectrum of the time, Darwin and FitzRoy viewed primitive peoples in the same way. They were both enthusiastic about the role of the missionary, and this was reinforced later in the voyage, in Tahiti and at the famous settlement at Waimate in New Zealand, where Darwin found ample evidence that the work of missionaries was beneficial to the indigenous people. Although Darwin and FitzRoy agreed about the role of the European in the civilising of primitive peoples, they disagreed over the reasons for doing it. Darwin, following the traditional Whig idealism of his family, believed in the equality of all humans and saw the work of the missionaries as a means to educate the less advanced peoples of the world. FitzRoy, with his orthodox Tory, Christian ethics, saw the role of Britain as a paternal force whose job it was to bring primitive races to a high enough level of civilisation to enable them to trade and thereby help expand the British Empire. Where the two men met in terms of idealogy was in agreeing about British philanthropy, where they differed was in the reasons behind it.

Although the *Beagle* did not stay in the Falklands for long, because of the strategic importance of the region, FitzRoy's work there was considered particularly important, constituting an important part of the surveying mission. The British had only recently acquired the Islands in the name of the Crown and planted the Union Jack there and the Admiralty wanted to produce accurate maps and charts of the region for both the merchant fleet and the Royal Navy.

The Falklands archipelago was a relatively uninspiring place for the naturalist but it did give Darwin a chance to pursue one of his early fascinations – marine creatures. While FitzRoy carried out his surveying work, Charles collected a variety of unusual marine and plant specimens to send back home in the next shipment to Henslow.

During the short stop-over, FitzRoy bought a small ship, which he paid for out of his own pocket. It was a small schooner which he saw as providing a relief for the overloaded *Beagle*. Unfortunately, the boat needed extensive work and so they had to return to Bahia Blanca to refit her. Darwin was disappointed by the delay. He was anxious to get to the west coast of the continent and in particular to visit Chile so that he could travel inland to the Andes. The detour proved to be only the first of several unexpected delays. On the way back to Bahia Blanca, FitzRoy changed his mind again and plotted a new course, taking the *Beagle* much further north to Montevideo, halfway back to where they had first made port in South America, almost a year earlier.

It was because of these changes of plan and last-minute decisions to make additional unscheduled investigations that the voyage of the *Beagle*, which was originally intended to take little more than three years, stretched into a journey lasting very nearly five.

Historians have often wondered why, when it looked as if the project was going nowhere, Darwin did not simply leave the ship and return home or else continue on his own. The answer is complex. First, Darwin was only occasionally unhappy with the unexpected course changes. He was, after all, on a mission of scientific investigation and all the places the *Beagle* visited were of tremendous interest even if they were unpleasant, cold or dangerous. Second, he was in no real hurry. Although he had a great desire to see the west coast of the continent and to get on with his scientific investigations there, they could wait as there was plenty to do on the east coast. Third, he felt that he could not desert FitzRoy no matter how often plans were changed. They had become friends despite their political differences and their frequent disagreements. In later life Darwin filtered out of his accounts many of the rewarding experiences and stimulating conversations he enjoyed with FitzRoy and concentrated rather too much on the Captain's failings, his tempers and mood swings. In reality, throughout the five years they were together on the voyage the two men remained cooperative colleagues and close partners. Darwin fulfilled admirably the role required of him and for FitzRoy's part, he was a very professional sailor, a highly intelligent and charming companion and a very good conversationalist, able to hold his own on a wide range of subjects. He was also a gentleman who had powerful friends and he introduced Darwin to a number of important people who provided help. Most significantly, he also saw to it that Darwin could send his collected samples to England regularly free of charge as official Admiralty cargo, even though Darwin's presence on the *Beagle* was to have been entirely self-sponsored and the samples remained Darwin's private property.

Finally, despite the numerous course changes, continued seasickness, the storms and the cramped conditions, Darwin was not actually aboard the *Beagle* as much as one might imagine. His longest stretch at sea during the voyage was one of 47 consecutive days, and of the entire time, from December 1831 to October 1836, he spent a total of only 533 days at sea (about eighteen months). Most sea trips were between one and three weeks in duration, and at a number of points during the voyage he spent three or four months at a stretch off the ship, travelling inland.

This first serious detour and later unexpected course changes meant that Darwin stayed on the east coast of South America for over a year longer than expected. The *Beagle* left the Falklands early in April 1833 and did not round Cape Horn for the last time until 10 June 1834. Ironically, because Darwin spent so much of his time ashore during this period, it actually proved to be one of the most fulfilling parts of his time on the east coast.

Heading northward during the spring of 1833, progress was slow because of the newly acquired schooner. They stopped off at ports along the way – Bahia Blanca, Buenos Aires and

Montevideo, finally docking at Maldonado so that work on the schooner could begin. On the way they picked up their first mail for months.

The news was mixed. Charles discovered that Fanny's marriage to Biddulph was not going at all well and that she was regretting ever having broken her relationship with Charles. For Darwin, this news only made matters worse. Thousands of miles away from home, he could do nothing about it.

There was other, more sensational news. Bored and uninspired, Charles' brother Erasmus had almost precipitated a scandal by becoming involved with another Fanny (née Mackintosh), wife of their cousin Hensleigh Wedgwood, only months after her marriage. A nasty scene had been avoided only by the fact that Fanny had become pregnant (by her husband, Charles assumed). Then among the tittle-tattle came the sad news that during the same year as Hensleigh's marriage his sister Frances Wedgwood (Charles' cousin) had died aged 26.

There had been an election in which, Darwin learned from family news, old Uncle Jos Wedgwood had won the seat for Stoke-on-Trent for the Whigs, but the government was still in disarray with trouble from the extreme left. The middle ground, which Jos Wedgwood represented, was finding it hard to maintain a balanced political climate and the country seemed to be once again on the verge of social and political upheaval.

Meanwhile Charles' sisters were still talking about the possibility of their brother's becoming a clergyman upon his return. For much of his time away Darwin was happy to go along with this idea. In his letters home to his sisters he supports their suggestions and those of their father, encouraging them in the belief that, once the voyage was over, he would be only too happy to settle down in some rural parish and to sort through his papers and collections in his spare time while maintaining the none-to-stressful responsibilities of the village cleric. Darwin was not lying or fooling his family when he wrote these letters. The simple fact was that he had absolutely no idea at this time just how important his work would be, both for practical and theoretical science. He certainly had no notion of evolution or the part he would play in its development and was completely unaware of how highly Henslow and his colleagues regarded the samples he was sending home. Although he was faithfully sending off sections of his journal as he wrote them, and these were being passed around the family and cooed over by his relatives, he had no idea that his correspondence with Henslow and descriptions of his experiences would be received with interest by the scientific community. As far as he was concerned, he had a job of work to do and became obsessed with his efforts.

Charles was in his prime, fit, healthy and filled with a lust to discover. Each new experience brought with it a thousand and one questions, some he could answer there and then by experiment, others had to wait until he could consolidate his experiences and find time to devise a cohesive scheme to explain what he had witnessed. During the voyage he gave little thought to the longer-term importance of what he was doing and assumed that he would indeed lead a quiet life after it was all over.

Darwin's first excursion inland during the spring and summer of 1833, while the schooner was in dock having a new copper bottom fitted, was with FitzRoy. On 9 May, with the ships docked at Maldonado, the two of them, with a small team of guides and servants, headed out of town into the raw, hilly territory of the surrounding region. For a couple of weeks they lived the life of gauchos, camping under the stars and hunting their own food. It was a dangerous land, dotted with bandit camps, a place where Englishmen were an oddity; the bandits cared nothing for the origin and purposes of their victims. But the two gentlemen and their guides were well armed, well trained and careful.

Back in Maldonado two weeks later, FitzRoy took over responsibility for the ships and monitored the progress of the repairs while, after a brief stop in town to replenish his supplies, Darwin headed back into the wilderness with a newly acquired young assistant, the ship's sixteen-year-old odd-job man, Syms Covington. On their travels they caught, bought, skinned and preserved a vast range of animals. When they were in danger of overloading their mules they brought their finds back to a small, rented room in town. Darwin left Covington to prepare the specimens for him to dissect, searching through the stomachs of his finds to discover as much as he could about their diet, analysing bone structures, weighing, and making detailed notes on the anatomy, habits and behaviour of a wide range of fish, birds and mammals.

When repairs to the schooner were complete, FitzRoy renamed it HMS *Adventure* and towards the end of July it set sail with HMS *Beagle* for Rio Negro. Once there, Darwin decided to make the journey overland with Covington while FitzRoy headed off for Bahia Blanca to continue surveying the region.

This was probably his most dangerous excursion of the entire voyage. In the Bahia Blanca area the only law was that of the gun. A General Don Juan Manuel Ortez Rosas, a regional bandit-leader, had been employed by the Argentinian government to exterminate the Indians of the region and every traveller had to have his permission to pass through his claimed territory. Reaching Rosas' outpost on the Rio Colorado on 13 August, Darwin was taken to meet the General. The two men had a brief discussion during which Charles explained his reason for being in the area. The General displayed little understanding for why a fully grown man should spend his time on such a purpose, but nevertheless supplied Darwin with everything he needed and granted him and Covington free passage.

Although Darwin needed the General's permission to cross his land, he clearly despised the man. Rosas was a blood-thirsty tyrant, a man with whom no one argued. There was no escaping the fact that he had been employed by the Argentinian government to commit genocide. Darwin knew he was dealing with a mass murderer, but had no choice if he wanted to explore the region.

After two weeks travelling across the pampas, they made it safely to Bahia Blanca. They were ahead of the *Beagle* and Darwin decided to make a return visit to Punta Alta where he had found fossils almost exactly a year earlier. This time they discovered an entire skeleton the size of a horse which Darwin carefully prised from the cliff face, numbered and catalogued the individual bones and boxed them for shipment home. Little did he realise it at the time, but these discoveries were to prove to be some of the most important finds of the entire voyage and to elicit the greatest interest of Henslow and his colleagues in England.

After a brief reunion with FitzRoy and the *Beagle*, Darwin was off again. This time the two ships headed back up the coast to survey the River Plate north of the Bahia Blanca and Darwin pressed on alone into the north-west of the region, ending up in the tiny ramshackle town of Santa Fé, where he promptly collapsed with a fever.

The illness cut short Darwin's plans to continue north-west and he instead decided to head towards Buenos Aires, where he could receive medical treatment if he needed it. Before leaving for Santa Fé, he had sent Covington back to Punta Alta to search for more fossils and had arranged to meet him in Buenos Aires soon after.

The trip to Buenos Aires was a nightmare; travelling initially aboard a crowded sloop, he soon became sick of his tiny cabin and left the boat after a short distance and continued the journey in a hired canoe. Running a fever and forcing his way through mosquito-infested swampland, by the time he reached the city, Darwin was exhausted and felt worse than he had done in Santa Fé. To add to his troubles, he arrived just at the point where a threatened armed rebellion, lead by General Rosas, was about to spill over into bloody conflict. Just hours before Buenos Aires erupted into violence, Darwin managed to secure safe passage out of the city through the help of local European officials. Covington, who arrived soon after Darwin, almost never made it and Charles was able to get his assistant on the last vessel out of the port with him only because of his brief acquaintance with Rosas.

Darwin had never been so delighted to see the *Beagle* as on 3 November when he and Covington finally rejoined her in Montevideo. He was now desperate to head back south to finish their business in the South Atlantic and then to round the Cape. He longed for the west coast, not only because it provided the psychological boost he needed and a sense that they were at last on the homeward leg, but he had always expected it to be the more scientifically interesting part of the voyage. He wanted to leave behind this mish-mash of revolutionary madmen and aggressive primitives; Darwin called the east coast 'this stupid, unpicturesque side of the continent'.

Unfortunately, FitzRoy had other plans. He had not finished surveying the Plate and it was to be almost another two months before they could leave the area. Darwin had recovered from the fever but, for the time being at least, he no longer had the stomach for long trips inland. Tired of the place, he decided to spend his time aboard the *Beagle* cataloguing his finds, writing his journal and carefully packing his specimens before sending them back to grateful, receptive hands in England.

By 6 December 1833 FitzRoy was finally ready. At last, the two ships slipped anchor and headed back south along the Patagonian coast.

The return to Tierra del Fuego took almost three months, the journey broken by a short rest over the Christmas holiday in Port Desire and violent storms in the Straits of Magellan at the most southerly tip of the continent. When, in late February, they finally reached Woollya Cove, where the anglicised Fuegians had been left a year earlier, they found, not unexpectedly, that the mission had failed miserably. York Minster and his young wife Fuegia Basket had persuaded Jemmy to return with them to their home island, where he was promptly beaten up and robbed. A few days later and totally disheartened, Jemmy had returned to Woollya Cove with nothing left from his previous life. By the time the *Beagle* returned he had reverted to his former savage ways, had a pregnant wife who shared his hut and he had lost much of his ability to communicate with FitzRoy, Darwin and the others aboard the *Beagle*. It was a sorry end to the whole project and Darwin rather optimistically said of it: 'I do not now doubt that he [Jemmy] will be as happy as, perhaps happier than if he had never left his country.'

This single incident, more than almost anything that happened on Darwin's journey, created a lasting impression concerning the nature of human life. Most especially it reinforced Darwin's view that instinctive behaviour cut deep. Because Jemmy's people had been adapted to their environment for thousands of years, their instinctive behaviour could not be changed by an alien culture quite as easily as FitzRoy had imagined.

From Tierra del Fuego, the *Beagle* and *Adventure* set course for the Falkland Islands, where Darwin again encountered military action. A gang of gauchos lead by a criminal named Antonio Rivero had murdered some of the British living there and FitzRoy was obliged to use his men to capture the gang. After a brief skirmish, the criminals were rounded up and Rivero was clamped in irons in the ship's hold.

The rounding of the Cape was not far off but the *Beagle* had to make one more detour before Darwin could finally bid farewell to the eastern coast – an expedition along the Santa Cruz river where the Admiralty wanted soundings made. Darwin was delighted. The geology of the region was quite different to anything he had previously seen and he had his first tantalising glimpse of the Andes – one of the major attractions of the west coast.

At last, on 12 May 1834, the *Beagle* made for the Straits of Magellan at the tip of the continent. As they headed south into winter, the temperature plummeted and it began to snow heavily. The seas were mountainous and the ship hit one storm after another – the worst sea conditions Darwin had yet experienced.

Adventure had waited in the Falklands for mail to arrive before setting a course to join the *Beagle*. It caught up in early June and as Darwin lay feeling seasick in his bunk, ice forming opaque crusts on the skylight of his cabin and the snow falling in torrents outside, he read through letters from home, posted the previous Christmas.

Political news told of the Poor Law which had been amended in Parliament. According to latest reports, the recently created welfare system was not working and the government was facing a middle-class rebellion.

The parcels from home contained much appreciated reading matter, including novels by Walter Scott, and a collection of books by a female social reformer called Harriet Martineau, who wrote fiction based on the political turmoil and social problems of the time. They provided much welcome relief from the tedium of travel and the continued anguish of seasickness. After reading them Darwin passed them on to FitzRoy and the other officers.

The most important news came not from the gossip and

the ceaseless prattling and moralising of his pious sisters, the continuing escapades of his brother Erasmus, or the political wranglings, but from Henslow – an appreciation at last of all Darwin's efforts. According to his mentor back in Cambridge, the samples Darwin had been sending were the talk of scientific society and his correspondence with Henslow, detailing his discoveries, had been received with enthusiasm. What was more, Henslow had arranged for Darwin's fossils, in particular the giant sloths he had unearthed at Punta Alta, to be displayed before the Cambridge meeting of the British Association for the Advancement of Science earlier that year. This, Henslow claimed, would establish Darwin's name within the scientific community – a ready-made scientific reputation for when the voyager returned. Henslow then went on to beg him not to leave the voyage in any circumstances.

This was easy for Henslow to say, sitting as he was in his warm country home surrounded by his family and friends; but even then, as Darwin lay in his hammock rolling around in violent storms and enduring such intense seasickness that he could not stand or walk on deck for days on end, the enthusiasm and encouragement of his contemporaries and superiors at home was all he needed. Although it could do nothing to improve his physical conditions, the fact that his work was being so well received, combined with the fact that the *Beagle* had finally rounded the Cape, made him feel infinitely better.

The first stop on the west coast was the island of Chiloé, just over 1,000 miles north of the Horn. It was a cold, inhospitable place where it rained continuously during the winter. Darwin always remembered it with sadness because, soon after arriving there, one of his closest friends on board the *Beagle*, the Purser George Rowlett, died after contracting a tropical disease.

From Chiloé they headed for the Chilean city of Valparaiso. The two places could not have been more different. Valparaiso was the cultural centre of the country and boasted of its European heritage, considering itself a London or a Paris of the southern hemisphere. Indeed, it was the most European place Darwin had visited for almost three years and he made the most of it.

The city was swarming with rich Europeans, landed gentry who had emigrated in order to establish vast estates here where thousands of acres of land could be purchased for a tiny fraction of the cost of farmland in Europe. Others ran businesses in the city and built grand houses for little more than the price of a tiny cottage in England.

One of these wealthy colonialists was an old classmate of Darwin's who had been with him at Shrewsbury School – Richard Corfield. He lived in a fine suburban house at the edge of the city with the Andes and open country as a backdrop stretching into the hazy distance.

With Corfield, Darwin enjoyed exploring in a very different style to his earlier adventures with the gauchos. Here he travelled some of the way inland by carriage and stayed in the houses of rich merchants, dining in the evenings and carrying out his researches by day. Ironically, despite travelling in comfort, during one of these expeditions Darwin again fell ill. In September 1834, while exploring a gold-mine in the foothills of the Andes, he collapsed with a fever and had to be taken to Valparaiso.

It is unclear what was wrong with him but it must have been quite serious because he was bedridden for a month. Darwin seems to have suffered a form of food poisoning, from, he speculated, some bad country wine he had sampled in a mountain settlement. Considering the alien climates and the often severe conditions of his life during the five-year voyage, it is surprising that Darwin was not ill more often. But then, not quite 26, he was in his prime. It was not until much later in his life that he was afflicted by a succession of debilitating illnesses, the origin of which, some believe, could have been linked with one of the illnesses he experienced on the voyage.

As Darwin lay in bed with a fever, other problems had arisen. The Admiralty had finally found out about FitzRoy's purchase of *Adventure* and was not pleased. The Captain was exhausted by the trip and being reprimanded by his superiors was the final straw. FitzRoy appears to have had a breakdown. In a rage he resigned his commission and started making plans to return directly to England. 'Hot Coffee' had boiled over again. The whole voyage was suddenly in total disarray. When, from his sick-bed, Darwin heard of FitzRoy's collapse, he began to plan his own route home. As soon as he was well enough he would travel across the continent before boarding a ship home in Buenos Aires, a journey that he estimated would take him well over a year.

Then, as suddenly as the voyage had fallen apart, it seemed to pull itself together again. FitzRoy recovered his mental equilibrium and changed his mind about his resignation, accepting the Admiralty's reprimand and apologising. Shortly after, Darwin regained his physical health and FitzRoy managed to secure the confidence of the crew.

Early in November 1834 the *Beagle* headed back to Chiloé, where they spent almost four months surveying the region, before plotting a course north along the Chilean coast, arriving in the port of Valdivia in mid-February 1835. Valdivia was another large town in the same mould as Valparaiso and the officers and crew indulged themselves, hosting a party on board the *Beagle*. They imported a boat load of señoritas from the town and, as Darwin lightheartedly noted in his diary: 'bad weather compelled them to stay all night. [It was] a sore plague to everyone.'

Valdivia turned out to be more than a pleasure stop. A few mornings after the party, as the ships lay at anchor in the harbour, around ten o'clock in the morning, Darwin was ashore sitting on the forest floor at the harbour's edge when he felt the earth move. He tried to stand but his legs gave way. A massive earthquake had struck the west coast of the continent. Afterwards, Darwin reported in his journal:

> I felt only the earth tremble, but saw no consequences from it. Captain FitzRoy and the officers were at the town during the shock, and there the scene was most awful, for although the houses, from being built of wood, did not fall, yet they were so violently shaken that the boards creaked and rattled. The people rushed out of doors in the greatest alarm. I feel little doubt that it is these accompaniments which cause the horror of earthquakes.[8]

FitzRoy put the *Beagle* out to sea as soon as possible and headed for the nearest large town of Concepción, 200 miles north along the coast, fearful of what they would find there and with little idea of what help they could provide. Nothing could have prepared them for the devastation they saw. Although the vast majority of the population had survived, all habitation for many miles around had been totally destroyed. First the villages making up the community of Concepción had been hit by massive earth tremors; then the whole area had been deluged by a giant wave which had swept away hundreds of people, livestock and what had remained of the buildings. When the *Beagle* arrived they were met by total devastation and thousands of homeless, distraught people. There had been sporadic looting and the local powers were unable to control the criminals. Darwin recorded the scene in detail:

> I was landed on the island of Quiriquina. The major-domo of the estate quickly rode down to tell us the terrible news of the great earthquake of the 20th; 'that not a house in Concepción, or Talcahuano (the port) was standing; that seventy villages were destroyed; and that the great wave had almost washed away the ruins of Talcahuano.' Of this latter fact I soon saw abundant proof; the whole coast being strewed over with timber and furniture, as if a thousand great ships had been wrecked. Besides chairs, tables, bookshelves, etc in great numbers, there were several roofs of cottages, which had been drifted in an almost entire state.[9]

The scene was horrifying, but after the emotional impact had diminished, Darwin was able to rationalise and to study the effects of the earthquake on the local geology. As well as tracing its epicentre, he succeeded in pinpointing the cause of the earthquake to incipient volcanic activity beneath the city of Concepción. It made him realise, more than anything else he had ever experienced, just how frail and puny human beings were and just how insignificant are the works of humans compared with the power of the earth beneath our feet. Along with the biological and the other geological finds on his journey around the world, his first-hand experience of the Concepción earthquake laid the foundations of Darwin's all-important acceptance of the idea that geological and (later in his reasoning) evolutionary change takes a great deal of time; that nature does not operate within the insignificant time-scales of human lives and endeavours, but that the power of nature is not only evident in the awesome pyrotechnics of earthquakes and volcanic eruptions but also in the slow, irresistible influence it brings to bear on the evolution of species and geological formations.

It is important to realise that, although before the voyage Darwin was a rather unorthodox Christian, he was nonetheless a serious believer and held relatively conventional views about the relationship between God and humans and the place of humankind in the universal scheme of things. As we discuss in later chapters, his experiences on the voyage changed many of these.

After doing as much as they could, a week after arriving in the region, FitzRoy set the *Beagle* on a course back to Valparaiso, and Darwin began preparing for one of his longest and most ambitious expeditions overland – exploring the Andes with his friend Richard Corfield.

In Darwin's mind, crossing the Andes was always to be one of the great highlights of the voyage and it did not disappoint. Overcoming the extreme cold, he ascended to 13,000 feet and looked out over the continent, through the frosty air to the 22,000 foot tall, snow-capped Mount Tupungato and the hazy distance beyond. He felt incredibly inspired by the vista, writing in his journal:

> We spent the day on the summit, and I never enjoyed one more thoroughly. Chile, bounded by the Andes and the Pacific, was seen as a map. The pleasure from the scenery, in itself beautiful, was heightened by the many reflections which arose from the mere view of the Campana range with its lesser parallel ones, and of the broad valley of Quillota directly intersecting them.[10]

After visiting the town of Mendoza they threaded their way back across the Andes through the Uspallata Pass, where he stumbled upon a fossilised grove of trees. In his diary of 30 March he reported:

> In the central part of the range, at an elevation of about

7,000 feet, I observed on a bare slope some snow-white projecting columns. These were petrified trees, eleven being silicified, and about thirty to forty converted into coarsely-crystallised white calcareous spar. They were abruptly broken off, the upright stumps projecting a few feet above the ground. The trunks measured from three to five feet each in circumference.

FitzRoy meanwhile had set the *Beagle* on a course directly north heading for the Peruvian capital Lima and was moving slowly up the coast. Darwin returned to Corfield's house and then, after making his farewells, he set off on another inland expedition, this time to Coquimbo and on to Copiapó, travelling through the mountains collecting mule-loads of samples on the way before rejoining the ship which, by the end of June, had finished its work in the north and had travelled back down the coast to meet him.

After a brief return to Lima, where civil unrest cut short their intended stay and confined them to the ships for most of the time they stood at anchor, the *Beagle* and *Adventure* set sail for the Galápagos Islands, arriving at the closest island, Chatham, on 15 September.

Darwin's short visit to the islands was made famous in later years by virtue of his discoveries based on the nature of the different species of wildlife he found there (explained in Chapter 10). To many people, Darwin's voyage aboard the *Beagle* was to the Galápagos Islands and the rest of the five-year journey was an irrelevance. This, of course, is quite untrue, but demonstrates the importance of his findings on the islands and their significance in his development of the theory of evolution.

Darwin thought the Galápagos ugly and inhospitable, but a veritable gold-mine for both the naturalist and geologist. The climate made it a haven for reptiles of all varieties and the islands were home to dozens of species of lizard and turtle. The islands were also volcanic which, according to Darwin, gave much of the terrain the look of an industrial wasteland. It was baking hot, temperatures sometimes reaching 93°F in the shade, there was little fresh water, and the only human inhabitants were a group of 200 exiles deported from Ecuador living on one of the islands, Charles Island. Most importantly to Darwin, the islands were populated by large numbers of different species of finch and tortoise (the latter were the staple diet and general resource for the exiles). Although Darwin did not at the time realise the significance of the various species of both of these creatures, differing as they did from one island to another, later, when he began to analyse his samples and to piece together his findings, their importance became clear and acted as the experimental backbone of his theory.

They stayed in the Galápagos for five weeks before heading west towards Tahiti. This crossing, although over 3,000 miles, was the most pleasant sea journey of the entire voyage. The weather was fine and Darwin was even able to take part in the normal life of the ship because he was only rarely confined to his bed with seasickness.

Although the initial view of the islands did not inspire Darwin, when the ship arrived there, he quickly changed his mind. They spent almost two weeks on Tahiti and were made very welcome by the friendly natives, whom, although at the same level of technological advancement as the Fuegians, Darwin found to be far advanced in their behaviour – a further puzzle to add to the many questions over human development brewing in his mind.

From Tahiti they steered a course for New Zealand, which involved another three-week sea journey. By now Darwin was desperate to return home and spent almost the entire voyage writing up his reports on the latest stages of the journey.

New Zealand was a primitive land, and thanks to the wholesale genocide by the Europeans taking over the country, the indigenous natives, the Maoris, were by then almost extinct. Those not butchered by the colonialists were in the process of having their souls saved by missionaries who were having considerably more success than Matthews and FitzRoy had experienced in Tierra del Fuego.

Although the country was still largely unsettled and the surviving Maoris savage in the extreme, Darwin did little exploring and spent most of his time in genteel company, staying on farms with their 'well-dressed fields placed there as if by an enchanter's wand'. It was a country of obvious extremes and, even for such a liberal as Darwin, after so long on his travels he could not help but be swept up in a sentimental mood for what he saw as the great British way of life.

The cultural contrast remained when, thirteen days after leaving New Zealand on 30 December 1835, the *Beagle* sailed into a crowded Sydney Harbour.

Sydney was a boom town, uncouth, fiscally obsessed and growing at an astonishing pace. As with New Zealand, the natives, the aborigines, were being exterminated freely or forced into slavery, and Darwin found the place extremely distasteful. The uniform geology also failed to inspire and, after sailing south to spend some time in Tasmania and then travelling up the west coast, Darwin left Australia in March with very little good to say about the place, writing in his diary:

> Farewell Australia, you are a rising infant and doubtless some day will reign a great princess in the South; but you are too great and ambitious for affection, yet not great enough for respect; I leave your shores without sorrow or regret.

It took the *Beagle* a further six months to reach home. They made good speed to the Indian Ocean, where Darwin was fascinated and delighted by the coral reefs, especially those around the islands of Keeling and Cocos, believing them to rank among the most beautiful objects in the world. From there they sailed west, arriving in Mauritius three weeks later. During the journey Darwin wrote up his journal, detailing his observations of the coral reefs of the Indian Ocean. From Mauritius they sailed down the east coast of Africa and stopped briefly at Cape Town, where Darwin visited one of his heroes, the great astronomer Sir John Herschel, who lived a few miles outside the city.

From South Africa it should have been a direct journey home to England, but much to Darwin's annoyance and despite angry mutterings from the sea-weary crew, on 23 July FitzRoy decided that they needed to make one final detour – back to Bahia in Brazil in order to recheck the longitude measurements taken almost four years earlier.

Although Darwin had suspected since they had left Mauritius that the captain wanted to go back to Brazil on the homeward leg, it did not soften the blow. He was utterly exhausted with travelling and longed for the comforts of home, reporting in his diary: 'This zig-zag manner of proceeding is very grievous; it has put the finishing stroke to my feelings. I loathe, and abhor the sea.'

When FitzRoy finally set a course for England on 6 August 1836 Darwin breathed a great sigh of relief, but their problems were not over. The *Beagle* was hemmed in by storms and they could not leave until the 17th. Darwin busied himself with his collections and returned to the solace of his diary and his journal. Crossing the Equator on 21 August and the Tropic of Cancer on 9 September, they were at last well and truly on the final leg.

As the *Beagle* sailed through a final storm during the night of 2 October 1836, inching its way into Falmouth, Darwin could hardly contain his excitement at being within sight of home. The voyage had lasted four years and nine months, but he had aged far more than the years signified. He had left England as a young man fresh out of university with only a good upbringing, a deep-rooted enthusiasm and an amateur knowledge of geology and naturalism to his credit. He was returning with 1,383 pages of geology notes, 368 pages of zoology notes, a catalogue of 1,529 species in spirits and 3,907 labelled skins, bones and miscellaneous specimens, as well as a live baby tortoise from the Galápagos Islands. His diary amounted to 770 pages, and parts of his journals, sent ahead of him, had already been read by a number of scientists at home. He was returning to England a little older but a great deal wiser. With his discoveries and finds already being discussed within the scientific community and with the greatest respect both of his contemporaries and elders having been gained, as he returned to England that stormy autumn night, although he had not yet realised it himself, the real work was about to begin.

Notes

Unless otherwise specified, quotations in this chapter are from *Charles Darwin's Beagle Diary*, ed. Richard Keynes, Cambridge University Press, 1988.

1 *The Autobiography of Charles Darwin and Selected Letters*, 3 vols, ed. F. Darwin, J. Murray, London, 1887.
2 *The Correspondence of Charles Darwin*, 8 vols, ed. F. Burkhardt and S. Smith, Cambridge University Press, 1985–93, vol. I, p. 133.
5 Charles Darwin, *The Voyage of the Beagle*, Colburn, London, 1845.
6–10 Gavin De Beer, *Darwin's Journal, Bulletin of the British Museum, Historical Series* (1959). (© The Natural History Museum, London).

Chapter 4

The Gift of Time

Long before Charles Darwin's day, the pioneering evolutionists such as Erasmus Darwin and Lamarck had realised that one thing evolution required was time – lots and lots of time during which infinitesimal changes from one generation to the next could add up to produce different variations on the evolutionary theme. This was one of the first causes of conflict with the established Christian Church.

Archbishop James Ussher published his *Sacred Chronology* in 1620. This was regarded as a fine example of Biblical scholarship at the time, and in it Ussher calculated the generations recorded in the Bible back from the time of Jesus Christ to the time of Adam, concluding that the Creation had occurred in 4004 BC. John Lightfoot, Vice-Chancellor of the University of Cambridge, then declared that the final act of Creation, the precise moment when Adam himself was Created, had been at 9 a.m. on Sunday, 23 October 4004 BC.

To put Ussher in his historical context, remember that this claim was made at a time when Galileo (who lived from 1564 to 1642) was being persecuted by the Church in Rome for suggesting that the Earth moves round the Sun. So when Descartes published his ideas about the origin of the Earth by natural processes involving matter initially spread through space, he was careful (in 1644) to say that this matter had been set in motion by God, who had foreseen the end product. Descartes developed an elaborate idea of vortices, saying that the Earth did not move freely through space but was carried around the Sun in a kind of whirlwind of particles, so that it could be described as 'stationary' with respect to the whirlwind, and he could thereby avoid Galileo's fate.

This was the beginning of attempts to explain the origin of the Earth itself in terms of natural laws – the laws of physics. It would be many decades before those attempts really began to bear fruit; but, as in the case of evolution of life, it became increasingly clear that the 'evolution' of the Earth itself required much longer than the six thousand years or so provided by Archbishop Ussher's calculations.

The first step towards a modern theory of the origin of the Earth, based upon Newtonian mechanics and using Newton's law of gravity, came from William Whiston, Newton's successor in Cambridge, in 1696, nine years after the publication of Newton's great work, the *Principia*. Whiston's idea was that the Earth was created out of a comet, a swarm of dust particles that had collapsed under the pull of gravity to form a solid object. The Biblical Flood was explained as being the result of another comet passing close by the Earth and dumping a large amount of water onto its surface. At this stage thinkers such as Whiston were still being careful to make their scientific accounts of the origin of the Earth match at least the outline of the Biblical story of Creation. But this became harder and harder as more and more evidence about the long history of the Earth began to emerge from studies of the rocks themselves.

When natural philosophers began to look closely at the nature of rocks, in the seventeenth century, they soon realised that some kinds of rock (now known as sedimentary rocks) had been laid down in parallel layers, which shows that they were formed from material deposited under water, at the bottom of lakes or at the bottom of the oceans. Additional evidence that sedimentary rocks formed in this way came, as Arabic scholars had noticed six centuries before, from the fossil remains of sea creatures, found embedded in sedimentary rocks even from tall mountain ranges. And by the end of the seventeenth century it was becoming generally accepted that fossils really were the remains of once-living creatures.

Before the invention of the microscope it was not at all clear that fossils were the remains of living plants and animals. Indeed, originally the word 'fossil' itself meant any peculiar object found in the rocks, some of which looked like living creatures, and others which resembled geometrical shapes or pretty patterns. When nobody knew how life itself had formed, it was easy to accept that stony objects which looked like living things could also have been formed naturally, within the rocks.

Things began to change when the microscope was invented. One of the pioneering microscopists, Robert Hooke (1635–1703), made many observations of living organisms, and published the first great work on microscopy, the heavily illustrated *Micrographia*, in 1665. Hooke was a skilled draughtsman and superb instrument maker, and produced detailed drawings of many insects, including fleas and lice (all too readily available in the seventeenth century), which caused a sensation at the time. Samuel Pepys records, in his diary for 1665, how he sat up until two in the morning reading Hooke's book.

Hooke realised that the creatures he observed in such detail with the aid of his microscope were superbly fitted to their roles in life, but he interpreted this as evidence of God's design, rather than adaptation through evolution. But he also examined a piece of fossilised wood under the microscope, and described it in the *Micrographia* and in lectures to the Royal Society that were eventually published in 1705. He noted that the detailed structure of the fossil closely resembled that of charcoal or rotten wood, and concluded that it really was a piece of wood that had been turned to stone (petrified) through 'having lain in some place where it was well soaked with petrifying water' so that it had become impregnated with 'stony and earthy particles'.[1]

He also realised that some fossils might be produced when organisms were buried in soft material to make a hollow mould, rather like the impression made by a seal in wax, which then got filled in to make a cast of the original creature.

All of this is very close to the modern understanding of how fossils form. And having realised that ammonites were the remains of sea creatures resembling the modern nautilus, Hooke was also faced with the problem of explaining how they got to be in sediments now found high in the mountains. He concluded that 'parts which have been sea are now land' and that 'mountains have been turned into plains, and plains into mountains, and the like'.[2]

But how? How had the surface of the Earth been changed so dramatically? That was one of the key puzzles facing eighteenth-century scientists, and at first they tackled it by trying to make the evidence fit the Biblical story of the Flood.

A crucial step forward was made by the Danish anatomist Niels Steensen (1638–86), a contemporary of Hooke's who (like Linnaeus) is better known by the latinised version of his name, Nicolaus Steno. Steno realised the full significance of something that seems obvious once it is pointed out – if fossils such as ammonites are the remains of sea creatures that fell to the mud at the bottom of the sea when they died, then there were no rocks above the mud at that time. In other words, when geologists find a stack of different kinds of sedimentary (or other) rock, one on top of the other, they have a sequence of rocks laid down one after the other. If the rocks have not been disrupted (perhaps by earthquakes), the oldest strata will be at the bottom. The layers of rock, therefore, provide a slice through time, a record of changes that have occurred during the life of the Earth. Furthermore, said Steno, writing in 1669, sedimentary rocks must always be formed in flat layers, as silt settles at the bottom of the water. Where we find strata tilted at strange angles today, or even bent into curves, we are also seeing the effects of changes that have occurred since the rocks were formed – direct evidence of deformation caused by (presumably) the processes that turned plains into mountains, and mountains into plains.

In spite of presenting these impressive arguments in favour of the evolution of the Earth itself, Steno did not see any conflict between his scientific insights and his religion. Indeed, he became a priest in 1675, was ordained a bishop in 1677, and did no more scientific work for the rest of his short life.

An early attempt to combine Biblical and scientific ideas about the origin of the Earth was made by Thomas Burnet (1635–1715), whose book *The Sacred Theory of the Earth* appeared in several volumes during the 1680s. Burnet worked within the time-scale provided by Ussher, but picked up on an idea put forward by Descartes, who suggested that as the Earth cooled from primordial hot, star-like material it would have formed a solid crust surrounding a layer of water. Burnet suggested that the uniformly smooth outer shell of the original Earth was the Garden of Eden, and that the Flood occurred when continued cooling made this shell crack and collapse into the waters beneath. Pieces of the original crust, left sticking out of the water, would form the continents and mountains that we know today. In this way a natural process of cooling was invoked to explain the Flood, but in order to square this with the Biblical account of the Flood as a punishment for humanity, it had to be argued that God knew in advance that humankind would fall from grace, and set the natural processes that would lead to the Flood (and everything else) in motion at the time of the Creation. This led to some tortuous theological wrangling about free will that we are happy not to go into here.

Burnet's book was important because it encouraged people to think about how the present-day features of the Earth might have formed naturally. But it had one glaring deficiency – on this picture, it is actually *harder* to explain why you find fossil ammonites high in the mountains, since no sedimentary rocks could have formed before the Flood. Although attempts were made to patch up this kind of hypothesis, in the first half of the eighteenth century it became increasingly clear that the Biblical story of the Creation and the Flood could not, after all, be squared up with the record in the rocks. When the work of Linnaeus and others showed that all of the animals on Earth could not have spread out from a single location (either Noah's Ark or some mountainous island), the way was opened for ideas which made no reference to the Biblical catastrophe.

One of the first people to take up the challenge was Georges Buffon, mentioned in Chapter 2. In 1778 he published a book called *The Epochs of Nature*, in which, although carefully dividing the history of the Earth into seven stages to match the seven 'days' of Biblical Creation, he described that history solely in terms of natural processes.

Buffon assumed that the Earth had been formed from a globule of hot material knocked out of the surface of the Sun by the impact of a comet. We now know that this is completely impossible (according to present-day astronomical thinking, the

Earth and the other planets of the solar system formed at the same time as the Sun, from a primordial cloud of gas and dust that collapsed under the pull of gravity), but he was right in thinking that the early Earth was a ball of hot, molten rock.

In Buffon's chronology, as the Earth cooled it developed a crust of solid rock. Later, water condensed and created an ocean covering the entire globe. Life arose in the oceans, and sedimentary rocks, containing traces of those forms of life, were laid down beneath them. As the level of the oceans fell, land began to emerge, and life moved onto the land, which was at first warm enough for 'tropical' vegetation and animals to exist everywhere. Finally, as the Earth cooled still further, the polar regions froze and even temperate latitudes became unsuitable for the original forms of land life.

This broad outline is impressive enough, for someone writing more than two hundred years ago. But the great leap forward that Buffon made was in rejecting the Biblical time-scale for all of this activity to have occurred. He actually carried out some experiments with hot balls of iron, in order to find out how long it took for them to cool. The larger a ball was, he found, the longer it took to cool down; and from this evidence Buffon suggested in print that it might have taken the Earth some 75,000 years to cool from a molten ball of rock to its present temperature. This was more than ten times the age Ussher had calculated – but even here Buffon was being cautious (or perhaps tactful) about what he said in print. To his friends, in letters that were not published at the time, he admitted that this was much too low a figure.[3]

The idea of a retreating ocean uncovering successive layers of rock laid down during earlier epochs was taken up by several people, and became known as 'Neptunism'. The obvious difficulty of where all the water had gone to as the oceans retreated was brushed aside, and many details of the theory did help towards a better understanding of geology. For example, the constant retreat of the oceans allowed continuous erosion from the land to lay down new sediments which were uncovered and eroded in their turn. The idea was championed by the German Abraham Werner (1749–1817), who was responsible for the first successful classification of the rocks in terms of their age – he coined the terms 'Primary', 'Secondary' and 'Tertiary' for the main epochs of rock formation, which remain the main divisions recognised by geologists.

Werner's ideas were published in 1786, in a book called *A Short Classification and Description of the Different Rocks*. But there was already a rival hypothesis to Neptunism, which held that the Earth's surface had been shaped by volcanic and earthquake activity, and which became known, logically enough, as Vulcanism.

The Frenchman Jean Guettard (1715–86) realised in the 1750s that many of the mountains in central France have the characteristic cone shape of a volcano, even though there is no record of volcanic activity in the region during the course of human history. Werner believed that volcanic activity was simply the result of coal burning beneath the surface of the Earth, and that no force was capable of lifting up the surface of the Earth. But in the 1760s another French naturalist, Nicolas Desmarest (1725–1815), mapped the distribution of basalt rocks around the Massif Central of France and showed that their patterns resembled lava flows. He also suggested that sheets of basalt such as the Devil's Causeway, in Ireland, were volcanic in origin. But although he saw all this as evidence for the great age of the Earth, and an indication that details of the landscape were shaped by Vulcanism, even Desmarest did not argue that volcanic activity could have built up whole continents. That step was taken by the Scot James Hutton, who lived from 1726 to 1797, and was already 59 years old when he presented his ideas about the evolution of the Earth to the newly formed Royal Society of Edinburgh, in 1785.

Hutton epitomises the natural philosopher of the eighteenth century – a kindred spirit, in some ways, to Erasmus Darwin. Although a farmer early in his life, he invented a process for manufacturing the chemical salt ammoniac, and made enough money out of this to devote his time to the study of the natural world, and to attending scientific meetings. These studies led him to conclude that no great acts of violence (such as the Biblical Flood) were necessary to explain how the world got to be the way it is, but that, rather, the nature of mountains and plains, and all the rest, could be explained in terms of the same natural processes that are going on today, provided that there has been sufficient time for them to do their work. This is the essence of a school of thought that became known as 'uniformitarianism'; it is the opposite of the idea of 'catastrophism', which holds that the Earth has been shaped by periodic violent upheavals at different times in the past.

In fact, Hutton took this principle to its extreme expression. He didn't give *any* date for the Creation of the Earth, but argued for what would now be called a 'steady state' model, in which the Earth had always existed and always would exist, constantly being worn away by erosion and constantly being rebuilt by Vulcanism. Most of the land surface is made of sedimentary rocks, and there is no difficulty in explaining how these could have formed under water from the debris carried down into the oceans by the rivers. The novel feature of Hutton's hypothesis was that these underwater sediments were then raised up by the action of subterranean heat, which formed new continents that were eroded in their turn.

Of course, previous ideas about the formation of the continents had also involved uplift, in the form of catastrophic great earthquakes like the ones associated with the cracking of the Earth's crust in Burnet's account. Even people who didn't believe that it had all happened in a single catastrophic event still argued that the upheavals that had shaped mountains in the distant past were bigger than anything that could happen on Earth today. But the uplift suggested by Hutton would be accomplished by an accumulation of relatively small earthquakes, each no bigger than anything that has happened in historical times, and by volcanic activity like the volcanic activity we see today, bursting through the crust at weak points and spreading basaltic lavas to make yet more layers of rock. And all of this activity would provide plenty of opportunity for strata to get tilted and crumpled into odd shapes in the process.

This is what required the vast span of time that Hutton talked about. Given enough time, it does not matter how small the uplift provided by each tiny earthquake is, eventually plains (or even the sea bed) will be raised up to the height of the Himalayas. He argued that if there ever had been an original formation of the Earth, the origins were lost in the mists of time, and that as far as the geological record was concerned we find 'no vestige of a beginning, no prospect of an end'.[4]

By the kind of time-scales envisaged by his contemporaries, he was not so far wrong. Modern estimates for the age of the Earth place its formation (along with that of the Sun and the rest of the solar system) some four and a half billion years ago, almost a million times longer than Ussher's estimate. The fact that we can now identify rocks billions of years old, and have a rough idea of what happened as the Earth formed all that time ago, is a triumph for modern science, not an indictment of Hutton. And although violent processes may have accompanied the actual formation of the Earth, modern geologists agree that, for the vast bulk of the billions of years of Earth history, continents have been created and destroyed by processes very much like those Hutton envisaged, and essentially the same as processes that go on today.

But Hutton's ideas were not well received in what remained of his own lifetime. A combination of Catastrophism and Neptunism held sway at the time, and Uniformitarianism was dismissed in the early 1790s. In response to his critics Hutton wrote a two-volume book, *Theory of the Earth*, which set out his views and appeared in 1795, just two years before his death. He had planned a third volume, which was never completed; but a summary of the uniformitarian idea was published in 1802 by Hutton's friend John Playfair. That, too, made little impression. In the first two decades of the nineteenth century, Catastrophism, championed by (among others) Cuvier, was much more representative of what the scientific establishment thought about the evolution of the Earth.

Georges Cuvier was born in 1769, the son of a Swiss soldier, and spent his early years in a small French town. He showed great academic promise, and went on to be educated in Stuttgart. He had read some of Buffon's books as a child, and they were to be a major influence on his life. After leaving Stuttgart, his first job was as tutor to the children of a family in Normandy, where he had ample opportunity to study natural history, and took a particular interest in the creatures of the seashore. The post, in a quiet backwater of France, also kept him away from the worst troubles of the French Revolution. In 1795 Cuvier was appointed assistant to the Professor of Comparative Anatomy at the Natural History Museum in Paris, the reincarnation under the new regime of Buffon's Royal Garden.

By 1799 he had been appointed Professor of Natural History at the Collège de France. In 1800 he began publishing what became a five-volume work, *Lectures in Comparative Anatomy*, and in 1802 he became Professor at the Natural History Museum itself. An able administrator, Cuvier played a major role in organising the new Sorbonne, and for the last twenty years or so of his life (he died in 1832) he was probably the most influential biologist in the world.

Cuvier extended the Linnaean classification of animals, and set new standards with the detailed accuracy of his descriptions of different animals – comparative anatomy. His insight into the way the different parts of a living animal work together then enabled him to make a great breakthrough in interpreting and classifying fossil remains.

Cuvier demonstrated the importance of a correlated set of body parts by describing the differences between flesh-eating mammals and plant-eating animals. He pointed out that a flesh-eater must have the right kind of legs to run fast and catch its prey, the right kind of paws to hold on to its prey, the right kind of teeth to tear meat with, and so on. A plant-eater, on the other hand, has flat, grinding teeth, hooves instead of paws, and other distinctive features. So if, for example, you find a few remains of some long-dead creature, and those remains include grinding teeth, then you know that the animal had hooves, not paws, even if you cannot find the remains of its feet. In his *Lectures*, Cuvier went so far as to claim that by looking at a single bone an expert would be able to reconstruct the essential features of a whole animal.

By the time he published that claim, he had already put the theory to the test, in 1796. Shortly after he took up his first post in Paris, Cuvier had been asked to analyse some fossil remains that had been sent to Paris from Paraguay. He interpreted the large fossil bones as belonging to an unknown species of giant sloth, no longer found on Earth today, which he called *Megatherium*. From *Megatherium*, he went on to reconstruct other fossil vertebrates, using the anatomical rules that he had developed from his studies of species that are alive today. Two key features emerged from these studies. First, many creatures found in the fossil record, like *Megatherium*, had no living counterparts. They represented species that had become extinct. Second, the older the rocks in which fossil remains were found, the more those remains differed, by and large, from species alive today. Almost single-handedly, Cuvier founded the branch of science known as paleontology, establishing it as the zoology of the past.

But Cuvier did not interpret these discoveries as evidence of evolution. Instead, he held the view that whole fauna, not just individual species, had been repeatedly swept from the face of the Earth in great catastrophes. After each catastrophe, the Earth was repopulated, according to Cuvier, by new species. There was no need to invoke the hand of God intervening directly to cause each catastrophe or to repopulate the Earth with new species each time. Although many people did, in fact, interpret the ideas of Catastrophism in this way (and Cuvier himself at one time talked of a 'succession of special creations', but later changed his mind[5]), most scientists preferred to argue that the initial conditions had been set up by God, but that everything following the original Creation had then worked out in accordance with the laws set down in the beginning – in accordance with God's Plan. On this picture, the Biblical Flood was only the latest in the series of catastrophes.

Unlike Hutton, Cuvier saw the history of the Earth in terms of long periods of relative calm, punctuated by occasional great catastrophes. Instead of seeing geological evidence such as the deformed rock strata of the Alps as a sign of gradual uplift, he interpreted this as indicating that mountain ranges formed in single, violent convulsions, causing changes in sea level that wiped out many species of plants and animals.

Lamarck opposed Cuvier's ideas, and argued that no species ever went extinct, but instead evolved into another form. He was, indeed, the biological equivalent of a uniformitarian, arguing that an accumulation of tiny changes, generation by generation, was what turned one species into another. This led to the bitter wrangle with Cuvier that ended with Cuvier's demolition of Saint-Hilaire and Lamarckism in 1830. The very person who had made the study of fossils a science, and who had himself pointed out the resemblance between the extinct species *Anchitherium* and the modern horse (a resemblance later taken as clinching evidence in support of evolution at work) had brought progress towards a theory of evolution to a halt in France. But even Cuvier had realised that the Earth must be much older than Ussher's estimate, and though he never gave a precise estimate he did write of a succession of past ages covering hundreds of thousands of years.

The idea of sudden changes from one environment to another, with accompanying sudden changes in the flora and fauna of the Earth, was influential across Europe, including Britain. The geologist William Buckland (1784–1856), of the University of Oxford, specifically used the ideas of Cuvier to argue, in his inaugural lecture as Reader in 1820, that geology did not undermine the truth of the Biblical story of Creation and the Flood. But geologists and fossil hunters were finding increasing evidence of a progression of life from the older strata to the younger strata, whatever the explanation for the causes of these changes. In 1824 Buckland himself was the first person to describe what we now know as a dinosaur, which he called *Megalosaurus* (the name 'dinosaur' was not coined until 1841, by Richard Owen). William Smith (1769–1839), an English canal builder who became interested in fossils through his work, established that each layer of rock contains distinctive fossil types, which can be used to place the strata in order. And the order that emerged from a combination of geological and fossil studies in the first decades of the nineteenth century showed that in the oldest rocks only invertebrate species, such as trilobites, were to be found, while in successively younger rocks there were the fossils of fish, then reptiles, and then mammals. It was against this background that another Scottish geologist, Charles Lyell, was to produce a book that had a profound impact on Charles Darwin.

Lyell was born in 1797, the same year that Hutton had died. He came from a wealthy family, and his father was known in scientific circles as a botanist. Charles Lyell was intended for a career in law, and as the first step towards this profession he was sent to Oxford University, where he became interested in geology through attending Buckland's lectures. But he also read one of the first popular accounts to support Hutton's uniformitarian ideas,[6] and did not emerge from Oxford steeped in the catastrophist viewpoint. While he was still a student at Oxford, Lyell visited Europe to study the geology for himself. He graduated in 1819 and went to London to study law, but continued to make field trips to Europe as often as he could; on one of these trips he met Cuvier in France in 1823. Although Lyell was called to the Bar in 1825, he never practised seriously as a lawyer. Using the excuse of poor eyesight (admittedly a genuine handicap to anyone who had to pore over legal briefs by candlelight) and funded by his father, he turned increasingly to geology, which soon became his full-time occupation.

The next major influence on Lyell came from a book about volcanoes written by George Scrope, which appeared in the mid-1820s. Scrope had studied the volcanoes of Italy, and had witnessed first hand an eruption of Vesuvius in 1822. He also studied the extinct volcanoes of France that had so impressed Guettard and Desmarest, and in his book he presented clear evidence that the landscape had been shaped by volcanic activity followed by erosion over a long period of time. Lyell set off on yet another expedition, to see for himself, and was lucky enough to discover that paleontologists working in the Massif Central had recently completed a study of fossil remains found in river sediments high above the present-day river valleys, but buried under a layer of basalt. Lyell realised that in the time it would have taken for the river bed to have carved the valley below, there would have been ample time for the fossil species found in those layers to have become extinct and been replaced by others.

He also visited Sicily, where he saw the geological strata which show that the mountain of Etna has been built up from many layers of lava, one on top of the other, also over a very long period of time. He returned to London in February 1829 fired with determination to write a book that would, once and for all, make the case that all geological phenomena could be explained by the same natural processes that we observe today, operating over very long periods of time.

Lyell's great work *Principles of Geology* was given a subtitle which made his intentions clear: 'An attempt to explain the former changes in the Earth's surface by reference to causes now in operation'. The first volume was published in 1830, and by July of that year Lyell was off on his travels again, to the geologically interesting region of the Pyrenees and on southward into Spain. This volume alone established Lyell as a leading authority in geology, and he was appointed Professor of Geology at King's College, London, in 1831, the year that the second volume of *Principles* appeared – the same year in which Darwin set sail aboard the *Beagle*. He gave up the post in 1833, however (the year the third and final volume of *Principles* appeared), in order to devote more time to geological studies in the field.

Lyell travelled widely throughout Europe, and visited America in 1841 and 1845. His work successfully established Uniformitarianism as the guiding principle in geology, and he was knighted in 1848 and elevated to the peerage, as a Baron, in 1864. But although he lived to see the great debate following the publication of Darwin's *Origin of Species* (he died in 1875), Lyell was only a reluctant, and incomplete, convert to Darwin's theory of evolution (he was, after all, in his sixties by the time Darwin published his theory), and never accepted that the theory could be applied to human beings.

In the *Principles* itself, Lyell had argued that species are distinct and unchanging units, and he presented Lamarck's ideas only in order to knock them down. He realised that species were well fitted to their environments, and argued that when gradual changes build up to modify the environment that means that some species are bound to go extinct, to be replaced by new species that are better fitted to the new environment. But where did the new species come from? Lyell never made that clear in his masterpiece, or later, saying only that when some species went extinct, other species 'took their place by virtue of a causation, which was quite beyond our comprehension'.[7] This obvious flaw in his argument was taken up by Charles Darwin as he began his own detailed studies of the natural world.

Darwin took the first volume of the *Principles* with him when he set sail on the *Beagle*; the second volume caught up with him later in the voyage, and the third volume was waiting for him upon his return to England in 1836. Darwin gained two things directly from Lyell – the idea that nothing more than the natural processes we see today, operating over a sufficiently long span of time, is all that is required to explain the great changes that have occurred on Earth during its history; and the realisation that there really had been a *very* long span of time over which those small changes could accumulate. He also gained something indirectly through Lyell – an appreciation of Lamarck's work. Ironically, although Darwin had read his grandfather's *Zoonomia* earlier in his life, he was not familiar with the details of Erasmus Darwin's mechanism for evolution, although he must have been more aware than many of his contemporaries of the possibility of evolution. So he actually learned properly about the idea of evolution by a gradual accumulation of small changes, and the (incorrect) mechanism of inheritability of acquired characteristics, from the *Principles*.

Although Lyell described Lamarckism in the *Principles* in order to refute the hypothesis, he did explain Lamarck's views clearly and thoroughly. Darwin set out on the voyage still a Christian (having, indeed, just missed being set on the path to a career in the Church), and almost certainly a believer in Special Creation. The chance to think carefully about Lamarck's views on evolution, albeit presented secondhand, was a step towards his own theory. But the geological gift of a sufficient span of time for evolution to have occurred was to be crucial in developing Darwin's theory of evolution, and the influence of Lyell's *Principles* on his thinking was so great that he later wrote:

> I always feel as if my books came half out of Lyell's brain, and that I have never acknowledged this sufficiently ... I have always thought that the great merit of the Principles was that it altered the whole tone of one's mind.[8]

Those books, however, still lay far in the future as Darwin completed his great voyage of discovery and set up home in England.

Notes

1, 2 Quoted by David Young, *The Discovery of Evolution*, Cambridge University Press, 1992.

3, 4 See Peter Bowler, *Evolution*, University of California Press, Los Angeles, 1989, [second edition].

5, 7 Henry Osborn, *From the Greeks to Darwin*, Macmillan, New York, 1894.

6 R. Bakewell, *An Introduction to Geology*, J. Harding, London, 1813.

8 Letter cited by Jonathan Howard, *Darwin*, Oxford University Press, 1982.

Chapter 5

The Traveller Returned

By the time the *Beagle* had docked in Falmouth, Darwin was so keen to get home to see his family that he left early the next morning, Monday 3 October. Driving his horses hard and travelling almost without a break, he arrived late on Tuesday night to find The Mount in darkness. Letting himself in, he crept up to his old room without waking anybody and the first the family knew of his return was when he casually strolled into breakfast the next morning.

The girls thought that he looked thin and tired and proposed that he have a long rest at home before sorting out his new life. It seems all ideas about Charles entering the Church had been dropped. Both his sisters and his father could see that he had begun to establish a name for himself from his work on the *Beagle* and the whole matter of a quiet country parish was hardly mentioned. Although he envied the life-style of his academic friends and their country houses, devoted wives and limited responsibilities, Charles had long since decided that his path could not be the easy one offered up by his family. But, at the same time, he had absolutely no idea what he was going to do.

His first priority was to find homes for his collections. Surprisingly this was not as easy as he had imagined. After little more than a week at home and a brief visit to Maer, he was off to see Henslow in Cambridge to get his advice on the matter and to talk about the voyage. Henslow put him in touch with several naturalists and other scientists in London, but the introductions produced mixed results. Darwin was certainly well known within scientific circles and Henslow's contacts were delighted to meet him, but many of them, the zoologists especially, were already overwhelmed with specimens. The geologists were keen to take the South American rocks off his hands and a few of the more exotic animals were accepted in very small numbers, but even the largest and most highly regarded of the London naturalist establishments, the Zoological Society, could take nothing, their store rooms and galleries were already brimming over with creatures brought back by traders, émigrés and military men. As Darwin had suspected from Henslow's correspondence during the voyage, it was the fossils which generated the most enthusiastic response. Shortly after his return Charles was introduced to his long-standing mentor in geology, Charles Lyell, and through him he was introduced to the man who would become both a friend for a time and the most dedicated researcher into Darwin's fossil collection – Richard Owen.

When the two men met at the end of October 1836, Owen was already establishing a name for himself as one of the most highly regarded zoology experts and ambitious anatomists in England and had just been appointed Hunterian Professor at the Royal College of Surgeons in London. That year a new library and a museum 90 feet long was nearing completion in the college grounds in Lincoln's Inn Fields and from his rooms on the top floor of the old building, Owen could watch the new constructions nearing completion. Owen was placed in charge of dissecting any animals that died in the Zoological Gardens and had a particular interest in fossils. With the new museum and library, he also had the room to take many of Darwin's treasures. Darwin met him in his rooms at college and was suitably impressed by the facilities as well as the man's growing reputation. Darwin needed someone with Owen's obvious capabilities to take care of his specimens and to research their origins and felt confident enough to hand over everything Owen would accept.

With the matter of his specimens dealt with, the next question was: how and where was he to work? Even though he had found the placing of his collections difficult, he still felt inspired by the interest shown in his journal of the voyage, and was confident that, through his contacts, a suitable publisher would be found for it. He also had ideas for other books. A travel book about the journey perhaps, a zoological treatise and smaller projects covering his interest in corals and his numerous geological findings. There was endless scope for academic consolidation of his five-year voyage, but he needed to establish himself somewhere in order to get started. He could accept his family's fervent requests for him to stay in Shrewsbury; that option would at least be cheap, secure and quiet, but he knew instinctively that now was the time for him to assert his independence; he had been away for too long to return blindly to the family fold and the cloying cosiness of it all. Equally, Cambridge had lost some of its allure. It seemed too small, too far from where everything was really happening, too quiet and selfconscious for his newly expanded view of the world. London looked ideal except for the fact that it was dirty and overpopulated. But then, on the plus side, it offered more opportunities, it was close to the centre of activity, his brother Eras had lived in London for the best part of a decade and now mixed in what Charles initially believed to be the most interesting and entertaining circles. As well as all this, in London he would be near his collections. In the end it was an easy decision. Despite harbouring an intense dislike for the cramped city and his deep-rooted desire to live in a quiet retreat in the countryside, he knew that now was not the right time; all that could come later.

Eras was leading as leisurely a life as ever and had been seeing a lot of the lady novelist Harriet Martineau, now famous throughout Europe for her political fiction. Charles disliked her, finding her far too domineering a character. Forever in despair over Eras' roller-coaster love life, Dr Robert and the rest of the Darwin family disapproved of the liaison, a fact which Charles only exacerbated by writing letters home in which he gave exaggerated accounts of the couple's activities.

Eras, who had already enjoyed a whirlwind social life there for many years, provided Charles with a perfect entrée into London society and introduced him to the cream of artists, writers and philosophers of the day. He hosted regular dinner parties and insisted that his brother expand his circle of friends, get out and enjoy his life rather than stay at home, hiding from the smog with his nose constantly buried in books or dabbling with unctuous preserving fluids and bits of dead animal. Eras was a keen scientist himself and conducted his own experiments but he had settled on a different balance to Charles. For Eras, life was 90% enjoyment, 10% dabbling in intellectual matters. For Charles at this stage in his career, he sought the same ingredients but in opposite proportions.

Eventually they struck a happy balance. Charles stayed at his brother's house for long stretches and returned to the family and occasionally to Cambridge, but gradually accepted that he was going to establish himself in London and that he would make the most of it. All the while he was forging stronger links with the capital. In January 1837 he was invited to deliver a paper to the Geological Society of London on the subject of coastal uplift in Chile and a short time later he was elected a Fellow. By mid-March it was clear that Charles would have to start spending even more time in London. Concluding that he had exploited Eras' hospitality for long enough, he took new rooms close by, a few doors along Great Marlborough Street, at No. 36.

The period from Darwin's return to England and his move to the country in 1842 was the most intensely productive of his life and marks the era during which he began to adopt the public persona of the gentleman scientist. By the early 1840s he would have several successful books to his credit, and the basic structure of his theory of the evolution of species filling a collection of secret notebooks.

His publishing career had been set rolling in November 1836 when he signed a contract with the publishers Colburn for his first book, to be called *Journal of Researches*. The project had been instigated by FitzRoy, who had suggested that he and Darwin should collaborate on an account of the voyage even before the *Beagle* had reached home. To FitzRoy's face Darwin had shown keen interest in the project, but was actually rather disappointed because he had already considered the idea of writing his own book about the voyage. After accepting FitzRoy's invitation to join the *Beagle*, Darwin could hardly refuse, so he went along with it, hoping that the Captain's plans would come to nothing.

In the long term it all worked out very well for Darwin. FitzRoy conceived the idea of a single volume containing the findings of Captain Philip King, who had commanded the *Beagle* on a previous trip, Darwin's journal of the most recent voyage and his own writings about the journey. Soon after publication in 1839, it became clear that Darwin's own contribution was by far the most popular and within three months of its issue, Colburn published Darwin's part separately. It became something of a Victorian bestseller and is still selling today in shortened form under the title *Voyage of the Beagle*.

The writing of the original version of the book was not without its problems. To the surprise of his friends and colleagues, FitzRoy had married almost immediately after returning to England. During the five years aboard ship, he had not once mentioned his fiancée despite the fact that Darwin had been his constant dining companion. After returning to England FitzRoy continued working full-time as a naval officer and had little time to spare for writing his part of the book. Charles was unhappy about this. Although preoccupied with other projects both mainstream and clandestine, Darwin was by this time a full-time writer and scientist and he had composed the original drafts of his journal during the voyage; all he had to do was to tidy up his manuscript and add a few notes. To make matters worse, FitzRoy was a pedantic and meticulous worker who would have taken longer than Darwin to complete his share even if he had been able to devote his entire time to the project. Whereas Darwin had his third of the book edited and checked by the summer of 1837, FitzRoy held up publication a full eighteen months.

The two men also clashed over the content of the book. Darwin wrote a preface for the first edition in which he carelessly forgot to credit the officers of the *Beagle* and mentioned FitzRoy only in passing. Being a stickler for etiquette, FitzRoy was not amused and sent a swingeing letter to Darwin insisting that the preface be rewritten. Then, when the book was finally published at the end of May 1839, it must have been a very peculiar read. FitzRoy espoused Christian views verging on fundamentalism. He was a creationist at the time of the voyage and married a woman who was, if anything, even more extreme than he.

Within months of his marriage he had become a religious radical believing in a literal interpretation of the Bible. Consequently, within the same volume one could find the earliest works of the man who was to cause perhaps the most profound rift between science and the Church in recent times, nestled alongside the work of a creationist which included a treatise describing the Flood, entitled *Remarks with Reference to the Deluge*.

In 1837, with his contribution to the *Journal of Researches* in full flow, Darwin was keen to capitalise on his findings and experiences aboard the *Beagle* and started work on two other projects. The first of these was a book to be entitled the *Zoology* based on the experts' reports of his specimens from the voyage – basically, an anthology covering the entire spectrum of the animal kingdom. It was remarkably ambitious but Darwin was fortunate enough and sufficiently well connected to secure a government grant of £1,000 to cover the cost of hundreds of engravings for the book. Even so, the *Zoology* was a demanding and time-consuming effort which eventually took over six years to complete. The other, less ambitious title was *Coral Reefs*, a project which Darwin dipped in and out of at irregular intervals during the rest of his time in London. Finally published in May 1842, it was sandwiched between work on the *Zoology* and his secret writings on the origin and meaning of human existence.

These works helped create the public image of Darwin. Along with his talks at the Geological Society and his eventual acceptance of the position of Secretary of the Society in February 1838, his scientific work was entirely orthodox. Yet, unknown to anyone (at least until confiding in Lyell in 1842), in the privacy of his own mind and in the writings kept safely in his study, Darwin was delving into areas completely at odds with the established thinking of the day and already formulating the earliest version of his theory of evolution.

He began in a notebook in July 1837, writing the word *Zoonomia* on the title page to signify to himself that he was following in the pioneering footsteps of his grandfather Erasmus. In this and a succession of later notebooks, he recorded all the questions and answers he had about nature, human life, evolution and religion as well as more mundane but connected observations and ideas. Although Darwin only half realised it at the time, he was working towards a grand synthesis of all these ponderables. Gradually, during the following years, he began to unravel the mysteries which had been troubling him since the early stages of the *Beagle* voyage. At the time he was only dimly aware of the eventual significance of what he was thinking and writing about, but he was also utterly convinced that he should not discuss his views with anyone.

A casual observation of Darwin's behaviour over the theory of evolution, and the way it eventually had to be teased from him by the arrival of Wallace's similar work many years later, has led many people to view Darwin as being unnecessarily cautious. For many, the creator of the theory of evolution is seen as a doddery semi-invalid who was too scared of the establishment to allow his secret theories to be published. This is far from the truth.

First, many of Darwin's ideas about evolution were formulated in the years between his arrival back in England and his leaving London to settle with his family in Kent. During this time, between 1836 and 1842, Darwin was indeed ill on many occasions, but he was still young, socially active and involved with a wide range of scientific pursuits. It was in fact, by his own admission, the most creative time of his life. But far more important than the image of the Darwin who devised the theory is the false premise that, by not going public with his origin ideas, he was in some way shirking his responsibilities. Darwin's theory took many years to develop fully and although he had devised the outline of his theory concerning the origin of species during the late 1830s, at that time it was in no shape to publish or even to show his close friends. That aside, the most important reason for his secrecy was his realisation that his ideas were simply too volatile for the time.

During the 1830s Britain witnessed one of the most turbulent periods in its history. It was an era during which power in the land shifted from the old order of Tory nobility and peasant workers who had very few social rights to a system dominated by Whig politics and the rise of the middle classes who were taking over the country in the aftermath of the Industrial Revolution, a revolution which they had controlled. The government of the time was guided by the ideas of the Reverend Malthus, an academic economist who believed in the law of the jungle as applied to human society. From the application of Malthusian ideals was built the social constructs of free enterprise, the rise and rise of the middle classes and the gradual erosion of the nobility. In the Malthusian order, all men were on an equal footing and were expected to find their own route through the world. In its crudest sense, it was a world in which each individual had to live by their merits, a philosophy which supported the merchant, the striver and spared little time for the weak or the poor. As a result, all social undesirables were disposed of by deportation to the colonies, the workhouses began to overflow and the country was convulsed with one rebellion after another. The Whigs were under attack from both sides. From one direction came the disaffected Tories, who had lost the security and the status quo of the old order; from the other came the rising discontentment of the masses, the poor workers who wanted a greater say in how the country was run. Throughout the years Darwin stayed in the capital, Britain was rocked by one upheaval after another, there was frequent fighting on the streets of London, riots across the country, widespread strikes and even armed suppression. This was quite evidently not the environment into which such anti-establishment views as Darwin's could be expressed openly; those stoking the fires of rebellion, the intellectual anarchists and revolutionaries, would have corrupted Darwin's theory, portraying it as a godless philosophy from a man at the very heart of the establishment.

Darwin was also conscious of the fate of other historical scientific revolutionaries, Galileo, Bruno and others. He wanted neither the ridicule of his contemporaries nor the condemnation of the political establishment. By the 1830s the Church had lost much of the power and enforced respect it had enjoyed during Galileo's lifetime; although Darwin was living in Britain, where the Inquisition never did have the power it enjoyed in Italy or Spain, the Church was still a potent force and an establishment with which Darwin did not wish to tangle. Furthermore he knew that, because of the political turbulence of the time, if he were to discuss his theories and they were made public, he would be forced into a battle on two fronts; something he was not ready to contemplate.

It was undoubtedly a great burden for him. He longed to

discuss his ideas with other scientists but he knew of no one he could really trust who would understand his theories and analyse them with an open mind. Compared with the stance Darwin was adopting by the late 1830s, Henslow was an extreme conservative. His old mentor in Edinburgh, the outspoken Robert Grant, would have been receptive but would also have been the worst possible person in whom to confide. Grant was a natural revolutionary and partly brought about his own academic and social suicide by having no regard for whom he talked to and what he declared in public. Darwin's new friends at the Geological Society and Owen at the Zoological Society were all establishment figures who could never accept such unconventional notions as Darwin's early ideas on natural selection and the evolution of life. Perhaps Charles could have discussed such things with his brother Eras, but it appears he did not. This is most likely because Charles could not rely on Eras to keep the subject secret; he was all too likely to spill the beans while half-drunk at a dinner party somewhere.

It was to be some time before Darwin could bring himself to confide in a fellow scientist, and in the end he chose Charles Lyell. Unfortunately, when he did, his misery was compounded by Lyell's total rejection of his ideas. Early in 1842, while Lyell was touring the United States, Charles outlined his ideas in a letter. Initially Lyell was disappointed and slightly bemused by such an extreme theory. For some time he seemed to confuse what Darwin was saying with Lamarckism, a theory he had so strenuously rejected years earlier. Because of this, Lyell wrongly concluded that Darwin's views did not support his own geological theories and refused to accept them. Even when Darwin's work was finally published almost two decades later, Lyell never did come out in total support of it and continued to procrastinate over the matter until his death in 1875. Although they remained close colleagues and personal friends, in some respects, Lyell's early rejection of Darwin's concepts was the beginning of the end of what could have been a closer and more productive scientific association.

As well as being the most prolific time of Darwin's life, the period between his return to England and his leaving London was also the period in which he began to suffer a succession of debilitating illnesses which were to continue for the rest of his life. Despite the fact that there have been numerous theories which attempt to explain the cause of these illnesses, none of them fully explains the facts.

The first sign that Darwin was suffering from recurring symptoms was in the spring of 1837 and from then until the end of his life Darwin went through the entire gamut of symptoms. There were of course periods during which he felt well but these rarely lasted for more than a few months at a stretch and his continuing illness became more severe after moving away from London in 1842. Some claim that Darwin did show symptoms of odd illnesses before the voyage of the *Beagle* and suggest that his health problems started at an early age or may have been genetic in character. But most evidence points to the fact that he really began to experience symptom after symptom, sometimes severe enough to lay him up in bed for weeks at a time, only after he had returned to England.

The first symptoms were stomach cramps and headaches, but during the following years Darwin experienced skin disorders, bouts of eczema, rheumatoid pains, insomnia, odd body swellings and heart palpitations. At college Charles had suffered from a mouth infection. He had scarlet fever when he was nine and had been seriously ill in South America, but these were all pretty standard illnesses of the day. Scarlet fever was common in nineteenth-century England and considering the level of hygiene and medical knowledge of the day, passing through adolescence with nothing more serious than a mouth infection in fact demonstrates that Darwin must have had a hardy constitution. Furthermore, the illness in Chile was an aberration and was one of the few times he was ill during the entire five-year voyage.

Theories concerning Darwin's health abound and new ones still appear at regular intervals. They range from suggestions that the death of his mother during his childhood caused a collection of psychosomatic symptoms in adulthood, to links between his later health problems and a tropical disease contracted during the voyage of the *Beagle*.

The possible connection between Darwin's symptoms and

his traumatised childhood was first suggested by the late John Bowlby, who was both a psychologist and a Darwin biographer. His premise was that Darwin suppressed the grief he felt for the loss of his mother and that this manifested itself later in adulthood as a series of physical symptoms.

The idea that early emotional trauma can result in health problems later in life has been well documented, but does not really hold water in Darwin's case. As recounted in Chapter 1, he was undoubtedly upset by his mother's death, but Charles was never really that close to her. Children of his class in the early nineteenth century were brought up by nannies and he rarely saw his mother. Furthermore, the same nanny stayed with the Darwins throughout Charles' early life and even sent messages to him via his sisters during the voyage of the *Beagle*. Darwin kept in touch with her and visited her in 1843 after she had retired.

Another strongly argued theory is that Darwin's health problems stemmed from the internal conflict over his secret work throughout the late 1830s and early 1840s. This matter did undoubtedly create huge stress for him. On the one hand, he was keen to be seen as a great scientist and to make what he knew would be an important contribution to human understanding. On the other, he was constantly frustrated by the fact that by revealing his theories he would be slated by his critics and his ideas corrupted by political extremists. He was in fear of having his reputation destroyed but also had a strong natural desire to have his work lionised.

The actual substance of his theories was another source of stress. Although he had been raised in a family still dominated by his grandfather's unorthodox Christianity, Darwin's upbringing had nonetheless been a Christian one. In developing his theory of the origin of species and a brand of humanist agnosticism, he was flying in the face of traditional values. Later his work would devastate accepted ideas of Biblical correctness and once more challenge the egocentric absurdities surrounding the place of human beings in the universal scheme of things. Despite the fact that Darwin's trust of science was far stronger than his belief in traditional religion, his conclusions were still hard to come to terms with.

It is clear that this stress and frustration greatly contributed to Darwin's health problems. Perhaps the most compelling evidence to support this comes from the fact that his symptoms began to appear in earnest at almost exactly the same time as he was formulating his unorthodox theories. Yet, some commentators strongly disagree. They claim that his illnesses developed before this time and that in fact there is a *lack* of synchronicity here which actually weakens the argument that the symptoms arose from the stress of his work. Critics of the stress theory point out that Darwin was ill several times before his return to England and the beginnings of his secret researches. But, as we have already suggested, these illnesses were either standard complaints of the day or extraordinary illness brought on by a particular foreign environment, such as his collapse in Valparaiso in 1834. So, what are we to conclude about the root causes of Darwin's health problems?

Perhaps the clearest description comes from combining accepted notions. Darwin's health was undoubtedly affected by the stress he was experiencing at the time the symptoms began to appear, but there must have been something wrong with him in the first place to precipitate the problem. Fabienne Smith, a writer who has made an intimate study of Darwin's health and written several learned papers on the matter,[1] comes to the conclusion that Darwin suffered from multiple allergies. The evidence is compelling. At the time, no doctor, not even Charles' own father, could provide a satisfactory diagnosis for Darwin's constant and various complaints. But, as Smith points out, Darwin suffered from hypersensitivity to heat, he came from a line of obese men (typical of multiple allergy sufferers) and Robert Darwin was allergic to some foods (especially cheese). She also points to the fact that Darwin's health seemed particularly vulnerable when he was using preservative chemicals, that gas light had recently been introduced in London and that he suffered most when there and later when conducting experiments at Down House. She also notes that both his brother Eras and Charles' own children appear to have suffered from an array of allergic responses, pointing to a possible genetic predisposition towards immune dysfunction. Finally, Smith suggests that the circumstances surrounding the death of Darwin's daughter Annie, at the age of ten, points to a total collapse of her immune system.

All of these factors, Smith believes, are evidence that Darwin suffered from a vast range of allergies and that it was this which produced the symptoms of nausea, headaches and heart palpitations which troubled him throughout his career. Coupled with high levels of stress, these environmental and perhaps genetic predispositions could explain the constant round of sickness and low spirits which prevented Darwin from working for long periods and forced him to spend an average of only two or three hours each day at his researches.

Another cause of Darwin's maladies, proposed by Dr Saul Adler during the 1950s and supported by no less a figure than Sir Peter Medawar, is that he contracted a tropical disease during the voyage which caused his collapse in Valparaiso and further manifested itself after his return to England. Their hypothesis is that Darwin was bitten by *Triatoma infestans*, the Benchuca bug which carries Chagas' disease. The argument still rages as to whether or not this disease lay at the root of the problem. The strongest aspect of this theory is that Darwin's symptoms appeared so soon after returning home and that, apart from some very common complaints, he was relatively fit and well before and during most of the *Beagle* voyage. But critics point to the fact that many of the normal symptoms of Chagas' disease do not fit medical reports of Darwin's symptoms, so the matter is by no means clear cut.

Finally, it has been suggested recently that Darwin suffered from myalgic encephalomyelitis (ME) and indeed, many of the symptoms he displayed do seem to fit the standard pattern of the illness. ME is closely related to high levels of stress and often attacks seemingly healthy people in their twenties and thirties. Without proper care and even with close attention it can last a lifetime; it produces symptoms of weariness and a whole catalogue of physical complaints. In fact, in many cases the disparate symptoms of ME are similar to those suffered by those with multiple allergic response and stem from the same source – a breakdown of the immune system. Sadly, we are unlikely ever to know the precise source of Darwin's ill-health except to conclude that it was probably related to an immunity dysfunction exacerbated by stress and that he may well have been weakened by contracting a tropical illness during the voyage of the *Beagle*.

Despite these constant bouts of illness, Darwin persevered with his dual pursuits. In his own opinion he was not producing as much as he would have liked and often described periods spent writing a pile of notes or organising his thoughts in a cohesive manner as wasted time. He was by no means single-minded. Despite frequent sickness, he continued to socialise and to work on his many projects, but he was already learning to pace himself. He was also keenly aware that he needed to sort out his personal life and to make some serious decisions about his future. In particular, there was the question of marriage.

Charles had ambivalent feelings on the subject. He wanted the security marriage could provide, he wanted to be looked after so he could concentrate on his work and not have to worry about domestic affairs. He also wanted to have children and was fearful of being alone in old age, secretly envying those of his contemporaries and peers who led quiet happy lives with families, pursuing their intellectual travels while enjoying the solid foundations of domestic comfort. In stark contrast, he looked upon his brother's life-style with little enthusiasm. Eras seemed to be enjoying bachelorhood, but it was not for Charles. Eras was altogether more gregarious and playful, less obsessed with science, and he had not benefited from the singular experience of the *Beagle* voyage. It might be true to say that, had Charles not taken his five-year sojourn from England, he could well have turned out a great deal more like his older brother.

As it was, by the summer of 1838, a few months short of his 30th birthday, Charles was giving a great deal of thought to the subject of marriage. In his usual analytical fashion he drew up a list of pros and cons to assess the situation. His biggest concern was that marriage would stifle him, prevent him from travelling if he decided he wanted to, that it would hinder his work by occupying too much of his time and that children might disturb his peace. It was an entirely selfish list of good and bad points, with scant concern for love or emotion; a purely scientific, pre-experimental treatment.

After a solitary walking and hammering trip to Scotland in June, Charles decided to talk over the question of marriage with his father and found Dr Robert to be entirely in favour of the idea. The real problem, Charles realised, was that there seemed to be very few suitable young ladies within his social circle. He was not keen on many of the women he knew in London; these were mostly Eras' friends and, with the exception of his first love, Fanny, now unhappily married, he had spent his entire life mixing with men. However, there was one young woman whom he had known since childhood, who was pretty and intelligent and whom he had encountered frequently at parties and dinners in London since his return from the voyage – his cousin Emma Wedgwood.

Emma was Josiah II's daughter and a sister of Hensleigh Wedgwood (who had been a close friend of Eras' until the latter had almost created a public scandal by becoming involved with Hensleigh's wife, Fanny). With family commitments stopping her from marrying long before, Emma was now just about the only eligible woman left in Darwin's immediate social circle and, considering the closeness of the Darwins and the Wedgwoods, it was inevitable that the two of them would have become good friends and possible partners. Emma had devoted most of her adult years to nursing her sick mother, a task she shared with her older, unmarried sister Elizabeth. It was this rather than any fault of Emma's which had kept her on the shelf. A little like a character from a Thackeray novel, chaste, lively and intelligent but not intellectual, devoted to her family but also realising that she had to lead her own life, Emma was the inevitable and, as time would prove, perfect partner for Charles.

Deciding that Emma was his best hope, Charles started to spend an increasing amount of time at Maer during the summer and early autumn of 1838. Despite his shyness and gentlemanly demeanour, it was obvious to all observers and indeed to Emma herself that they were courting. Yet, for a number of reasons, it took Charles four months to propose.

First, he wanted to be sure that he was making the right decision. He needed to know what Emma felt about him and about the whole issue of marriage before he could dip a toe in the water. Second, he considered himself to be so plain in appearance that she would not be interested in him and was afraid to suggest marriage for fear of rejection on the grounds of his ugliness. Third, and most important, during this period he was working intensely on the religious and philosophical ramifications of his ideas about the origin of species. He felt torn between keeping his ideas secret from Emma, which involved not knowing how she would react if ever they became public, and longing to talk to her about his work in order to clarify the matter before their relationship grew too serious. It was an agonising decision for him and in the end he decided simply to drop ideas into their conversations and see what the response would be.

Emma had a simplistic, orthodox faith, typical of her class during the early Victorian era. She believed that one must accept the teachings of Christianity in order to save one's immortal soul, that non-believers would be cast into the pits of hell. When Charles revealed to her a chink of light from the lantern of his ideas concerning the place of humanity and the hugely diminished role of God in his scheme of things, she was genuinely horrified that Charles was running a grievous risk, that his soul would suffer and that when they died, they would not enjoy eternal life together.

They discussed the problem, but it was difficult for Emma to explain her views fully because Charles always had the habit of reducing everything to its fundamentals, of parrying all arguments with cold scientific logic. So instead she wrote him letters, a habit which persisted throughout their lives together. In these letters she could pour her heart out and describe her feelings without clashing over meaning. Charles could then mull over her arguments before talking to her again. One of these letters (written soon after their marriage), in which Emma expresses her anxieties and tries to make her husband see the matter from a more emotional, spiritual aspect, affected him so much that he was moved to tears. A note that he added to the bottom of the letter reads: 'When I am dead, know that many times, I have kissed and cryed over this.'[2]

Charles proposed to Emma in November 1838. Having fallen

in love with him during their frequent meetings and deeply personal conversations, she could only hope that Charles would come to further appreciate her views and move back to a form of Christianity. Emma was still fearful for the preservation of Charles' soul, but she happily accepted him. Once over this hurdle Charles had little interest in the mechanics or the ceremony of marriage and wanted to have the entire matter behind him as quickly as possible. More practically minded, Emma realised that they would need time to arrange things and eventually they agreed that they should marry early in the New Year and set the day for late the following January.

There was much to do. While Emma and the families got on with the business of organising the wedding, Charles began house-hunting in London. He eventually found a house which Emma approved of in Upper Gower Street, Bloomsbury, close to the newly established 'godless' University of London. It was gaudily furnished but the furnishings and fittings came cheap. Ever cautious with his pennies, Charles was particularly pleased because the rent was very low.

Christmas and New Year was an exciting time with both families totally preoccupied with the forthcoming wedding. But, if December was busy, January 1839 was almost overwhelming. On the 24th, a few weeks before his 30th birthday, Charles was made a Fellow of the Royal Society. Before learning of this honour, the original date for the wedding had also been 24 January. Fortunately, by coincidence, because of complications, the Wedgwoods had been forced to put the date forward to the 29th. Charles had been annoyed about this until he discovered that if things had stayed as they were, he would have had to have been in two places at once.

In the event, Charles was almost casual about the wedding day itself. To him it was perhaps a rather silly ceremony, steeped in tired, anachronistic tradition, and he appears to have shown little regard for the feelings of Emma or the two families. The service was Anglican but specially altered so as not to offend the large contingent of Unitarians in the congregation; they had chosen another cousin, John Allen Wedgwood, to marry them. Surprisingly, there was no proper reception. Instead, Charles whisked Emma off to the railway station with almost indecent haste and in so doing antagonised a number of relatives from both families, especially Emma's closest friend and sister Elizabeth, who was now left on her own at Maer to look after their mother.

The couple named their house Macaw Cottage because of the colour scheme and immediately had the place redecorated and the hideous furnishings replaced. Once the aesthetics had been attended to, the house turned out to be ideal for their needs. They were at the very centre of activity, close to amenities, but their immediate area was quiet and the study was a silent retreat from the hurly-burly of city life. It was a large house and provided them with plenty of space in which to raise a family. A team of servants and maids moved in with them and Charles quickly returned to his labours, leaving the unpacking and arranging of furniture to his wife and the servants while he got on with his work.

Syms Covington, Darwin's right-hand man from the voyage, had been kept on as his manservant after the *Beagle*'s return and he moved into the Upper Gower Street house with Charles and Emma for a short time, helping them to settle in. Bright and ambitious, Covington did not want to spend his life in service. He left the Darwin home in February with a £2 golden handshake from Charles and his heart set on returning to Australia where he later became a clerk.

Emma became pregnant with their first child at the beginning of April and they started retreating from the social round of the London middle classes and began to adopt the mantle of an introverted couple content with each other's company. Charles developed a routine and worked around his constant ill-health. With domestic responsibilities left in the hands of servants, Emma was able to nurse Charles when he was ill and they grew emotionally closer. She continued in her attempts to mellow Charles' clinical views concerning the meaning of human existence and even managed to persuade him to accompany her to King's College Church on the Strand most Sundays. For his part, Charles was genuinely in love with Emma and in all likelihood went along with the ritual simply to keep her happy. She in turn gave her husband the moral support he needed to persist with his work, although feeling intuitively that he was treading on dangerous ground.

The Darwins were, for the most part, happy in Upper Gower Street. Charles continued to yearn for the countryside, continued to agonise over the social and religious implications of his work on evolution, and maintained his struggle against recurring illness. Despite the fact that he suffered some of his worst bouts of sickness, migraine and skin irritations during this period, within four years he fathered their first three children, wrote *Coral Reefs*, edited most of *Zoology* and filled notebook after notebook with his unpublished, private researches. In his autobiography, he declares: 'Whilst we resided in London, I did less scientific work, though I worked as hard as I possibly could, than during any other equal length of time in my life.'[3]

When their first son William was born on 27 December, their happiness was complete. Charles was absolutely delighted to be a father and was quick to realise that he had also been provided with a perfect specimen to study. During the first few months of the child's development, throughout early 1840, William Darwin was the constant subject of Charles' researches into the behavioural patterns of infant humans, providing further material in support of his father's theory of evolution.

By contrast, the remainder of 1840 was a bad time for Charles and turned out to be the beginning of one his most unhealthy periods, forcing him to lead an almost reclusive existence in Macaw Cottage. In March he attempted to resign as Secretary of the Geological Society but was persuaded to stay. Feeling too ill most of the time even to make the short trip to meetings, he decided that he would make it clear by his absence that he could no longer face the responsibility of the post. Eventually, the following spring, his resignation was accepted.

There were other reasons for his growing disillusionment with the Society. The real problem lay in the fact that he could no longer cope with the hypocrisy of what he was doing. The Geological Society was extremely orthodox and supported traditional scientific views, yet at the same time he was privately developing theories which went way beyond Lyell's. His theory of evolution and its links with geology would, in the long run, entirely destroy the old order and topple the edifice of unthinking Christianity as well as many of the accepted scientific notions of the day. Yet, as Secretary, he was supposed to uphold the values of the Society; he had even sat in silence at a Society meeting in December 1838 when Robert Grant had his (albeit ridiculous) views on evolution systematically pilloried by the likes of scientific conservatives such as Richard Owen.

In the political world, things were going from bad to worse. By 1840 radical reformers were demonstrating almost daily on the streets of London and the university towns. The country was in a deep depression, the newly crowned Queen Victoria, barely out of her teens and on the throne for less than three years, ruled a country on the verge of global empire, but witness to near chaos within its own shores. The Malthusian ideology was meeting with determined resistance from the working classes, who were being organised by intellectuals aiming to see a downfall of the class system, a fairly elected House of Commons, yearly elections and the right of each citizen to say who ran the country. It was a long way from twentieth-century socialism, but the supporters of the People's Charter, Chartists, as they were dubbed, could be seen as the precursors of the Labour Party and the Trade Union movement.

London was becoming a dangerous place in which to live and Charles was acutely aware of the situation. Staying at The Mount alone during the summer of 1841 to recover from a particularly serious attack of his usual symptoms, he approached his father with the idea that he might buy Charles and Emma a house in the country. Charles found his father surprisingly willing and upon returning to London the couple started house-hunting again, this time in Kent.

It took them until the following July to find the home in which they would spend the rest of their lives, a former parsonage called Down House, in the village of Down near Farnborough. Emma was not certain at first and thought the North Downs desolate but, on a later visit on 22 July, they stayed overnight in the public house in the village and found the locals warm and friendly and the village itself quiet and secluded. For Charles it was ideal: large and remote from the madness of the capital, but close to a village and communications with London if they were needed. What was more, at £2,000 it was cheap.

Returning to London, they could not have found the contrast between their prospective new home and the violent atmosphere more pronounced. On 14 August and for the following two days, the Guards and Royal Horse Artillery were on the streets of central London quelling the increasingly violent crowds. From their windows the Darwins sat watching the battles between troops and demonstrators merely yards from their front door. There was a palpable sense of impending doom hanging in the smoggy air, a genuine terror that the army and the police would lose control and that the country would slide inexorably into civil war.

It was the final straw for Charles. His yearning for the clear air and the solitude of the shires had not been enough to break away, but there was more than enough reason to leave, with two children and a third on the way, his health as bad as it had ever been since moving to the capital and fears for the safety of his beloved wife. Within hours of returning to the capital, and even before witnessing the street fights and the charging horsemen, the Darwins had set their hearts on purchasing Down House and moving away as quickly as possible. On 14 September Emma took their son William and tiny daughter Annie to the new house and two days later Charles left Macaw Cottage for the last time. Even then, six weeks after making the decision to leave, chaos reigned still in the capital. Some of Darwin's friends and colleagues had joined the security forces, Richard Owen had started to drill with the Honourable Artillery Company and was being called upon to help out the overstretched police force.

Darwin was set on escape. If the country did erupt into all-out civil war, he would only be delaying the day when he and his family might be drawn into the conflict, but he could do little about that now. In the short term he could only think of building a better life for himself, his wife and young children. Although he had dreamed of this move since floating lonely and homesick along the South American coast, as Charles closed the door on his old London home, even he might not have imagined just how many of his dreams and ambitions would be fulfilled at Down House.

Notes

1 Fabienne Smith, 'Charles Darwin's Ill Health', and 'Charles Darwin's Health Problems: The Allergy Hypothesis', *Journal of the History of Biology*, 23/3 (Fall 1990), pp. 443–59, and 25/2 (Summer 1992), pp. 285–306.

2 *The Correspondence of Charles Darwin*, 8 vols, ed. F. Burkhardt and S. Smith, Cambridge University Press, 1985–93, vol. II, pp. 171–2.

3 *The Autobiography of Charles Darwin and Selected Letters*, ed. Francis Darwin, John Murray, London, 1887.

Chapter 6

Early Works

The name of Charles Darwin is now so inextricably linked with the theory of evolution that it comes as a surprise to learn that he first made his name in scientific circles, and to the educated public of Britain, as a geologist. Of course, the zoological and botanical samples that he had sent back to Britain during the voyage of the *Beagle*, and the specimens brought back by the *Beagle* herself, were of enormous importance at the time, and would alone have made his name in specialist circles. But in the 1830s geology was the most glamorous of sciences, occupying a place held by cosmology today, and for much the same reasons. It dealt with vast stretches of time, and the origin of the world as we know it.

Geology books – not just popularisations, but some scientific volumes – sometimes outsold the popular novels of the day, and Lyell's *Principles of Geology* had sold thousands of copies. Geologists of the day ('hammerers', as they were affectionately known from their compulsion to chip away at rocks) were the Stephen Hawkings of the early Victorian era, and Darwin was soon established among the pantheon, first as a disciple of Lyell and then, before long, as a new star to rival the brilliance of the master himself.

The groundwork for this success had been laid, ironically, when Darwin was neglecting his proper studies of medicine and divinity in Edinburgh and at Cambridge University. Although, as we have seen, he did waste some of his time with the hunting set, there was always a more serious side to his character, shown at an early age by his fascination with chemistry. Tell young Charles Darwin what to study, and the chances were that he would rebel against the good advice (at least at first); but if he found a genuine interest of his own, then he would pursue that interest with an unmatched intensity, working far harder at his hobbies than at his set tasks.

For Darwin, the road to geological fame began with the molluscs of the seashore near Edinburgh, and travelled via botanical studies in Cambridge. But this was not such a strange route to take then as it would be now; science was less compartmentalised than it is today, and someone interested in the world about him did not have to specialise at an early age as a marine biologist, or a botanist, or a geophysicist.

Darwin's first important unofficial mentor, Robert Grant, was a free-thinking radical who had given up his medical practice to study marine life and lecture on invertebrate anatomy in Edinburgh. Although born in Edinburgh, Grant had travelled widely on the continent, was familiar with Lamarck's ideas about evolution, and rejected the idea of a series of Special Creations. As seen in Chapter 1, Darwin and Grant met through the Plinian Society; they studied primitive marine organisms together, and Darwin made significant contributions to these investigations. He learned not only about the creatures they were studying, but the techniques of scientific investigation in general, and the skills of dissection and analysis of living things in particular.

He also heard ideas which, although he did not accept them at first, must have had a subconscious influence. In 1826, during Darwin's first year in Edinburgh, Robert Jameson, the founder of the Plinian Society, was the first person to use the word 'evolution' in its modern sense, in an anonymous scientific paper praising Lamarck. But there is no evidence that evolutionary ideas had much direct impact on Darwin at the time, although it is known that Grant was familiar with the work of Erasmus Darwin, and it does seem likely, as one biography has suggested,[1] that Grant's willingness to accept Charles Darwin as an unofficial pupil owed more than a little to the fame of his grandfather. As Darwin later commented in his autobiography, hearing the work of Erasmus praised during his time in Edinburgh must have had some influence on his own willingness to take the idea of evolution seriously a few years later, when he made a careful study of Lyell's outline of Lamarck's work while on board the *Beagle*; but he certainly did not accept his grandfather's ideas on evolution when he first came across them.

Jameson himself began Darwin's geological education. He taught a natural history course which was not part of the required curriculum for medical students, but which was understandably *de rigueur* for the young Plinians. A hardline Neptunist, Jameson taught that all rock strata had been formed by sedimentation in the primeval ocean. The theoretical aspects of the course were already becoming outmoded in the 1820s, but Darwin also learned about Vulcanism from the chemistry professor in Edinburgh, Thomas Hope. More importantly, like his fellow students, under Jameson Darwin also learned about the succession of the strata, and how to relate the different layers of rock to one another – and he began his career as a 'hammerer' proper, going off on field trips to bash away at the rocks, learning to identify the strata in the field.

There was a bonus. As a natural history course, Jameson's classes included tuition in preserving and describing flora and fauna. Although almost completely useless for anyone seriously intending to practise medicine, Darwin's unofficial, self-chosen Edinburgh education could hardly have been better designed to fit him to the role of ship's naturalist on a voyage around the world; and there was more to come when he moved to Cambridge.

Cambridge University had its origins as a religious institution, and even at the beginning of the nineteenth century the path to a college fellowship involved taking holy orders and forsaking marriage – even for scientists – although marriage was possible if a man took up a university post as professor. So it is no surprise that two of Darwin's chief influences during his time in Cambridge were both Reverend gentlemen, John Henslow and Adam Sedgwick. Although there was no longer an elder brother in Cambridge to look after Charles, he was taken under the wing of his second cousin, William Darwin Fox, who was already at Christ's College.* Fox showed Darwin the ropes, and introduced him both to the delights of beetle hunting and to Henslow, who had taught Eras mineralogy.

Henslow was by now Professor of Botany. Eras and William Fox agreed that he knew everything about everything scientific, and Charles duly attended his lectures in botany – not once, but in each of his three full years in Cambridge. On Friday evenings Henslow kept open house for anyone interested in science, and in due course Darwin became a fixture at these meetings, discussing a wide range of scientific topics and slowly building up a friendship with the professor.

Although at this time Darwin was much taken with Paley's argument from design, and saw the perfection of the living world as evidence of the hand of God at work, this did not stop him reading widely on scientific topics, especially during his final year in Cambridge. He was particularly impressed by a new book published in 1831 by John Herschel, the son of the astronomer who had discovered Uranus, in which he spelled out the principles of scientific investigation. And Henslow encouraged him to read Alexander von Humboldt's classic and lengthy (six-volume) description of his travels in South America. In his autobiography, Darwin wrote that between them these two books fired him with 'a burning zeal to add even the most humble contributions to the noble structure of Natural Science. No one or a dozen other books influenced me nearly so much.'

The immediate effect of reading Herschel and von Humboldt was to infect Darwin with a rather wild scheme to make his own expedition to the island of Tenerife, in the Canary Islands, taking Henslow along as well, if he could be persuaded. Whether this plan would ever have come to anything if the much larger opportunity of the *Beagle* voyage had not intervened, and whether Darwin would have benefited from the trip if it had come off, we shall never know. But the plan itself had one important and immediate effect. Darwin realised that if he was to do any serious

---

* In a delightful coincidence, during much of Darwin's time in Cambridge the rooms he occupied at Christ's were those where William Paley had lived as an undergraduate, a little over half a century earlier.

scientific work on Tenerife, even the most humble contribution, then he would need to know a lot more geology to go alongside his now considerable botanical expertise. He sought Henslow's help – and Henslow sent him to Sedgwick, the Professor of Geology, for some cramming in the subject.

Henslow himself was a former star pupil of Sedgwick who had very much made good, and his recommendation was enough to encourage Sedgwick to take Darwin seriously. Towards the end of his final year in Cambridge (after taking his final examinations, but while keeping the obligatory last two terms' residence to complete the qualification for his degree) Darwin attended Sedgwick's lectures, took tuition from the professor, and even tried to learn Spanish as a prelude to his proposed trip to the Canaries. Sedgwick was sufficiently impressed with Darwin's enthusiasm, growing expertise and personality that he invited the young man (now 22) to accompany him on a field trip in the summer, mapping out the complex geology of north Wales.

In three weeks, during the summer of 1831, Darwin had a crash course in practical geology from the hands of a master. Delighting in his new-found skills, and glowing with praise from Sedgwick for his valuable contributions to their studies, in August he left the master in north Wales (at Capel Curig) and set off cross-country on his own, heading in a straight line by compass through wild and rugged regions to meet up with friends at Barmouth, 30 miles away. He kept up his geological work along the way, hammering away, identifying and mapping the strata as he went.

By now, Darwin was a confident young man armed with a wealth of ideas about the natural world, backed up by considerable practical experience. As Darwin himself appreciated, he had become a true scientist on this expedition with Sedgwick:

> Nothing before had ever made me thoroughly realise, though I had read various scientific books, that science consists in grouping facts so that general laws or conclusions may be drawn from them.[2]

So the young man who returned to Shrewsbury at the end of August was now thoroughly versed in botany and entomology, had more than a smattering of zoology, knew how to prepare and preserve specimens, was an expert field geologist who had been taught his trade by one of the Vice-Presidents of the Geological Society of London, and had a modest working knowledge of the Spanish language. It has sometimes been suggested that Darwin was something of a wealthy playboy who got his berth on board the *Beagle* simply by being in the right place at the right time. To be sure, he *was* in the right place at the right time – but the reason why Henslow recommended him to FitzRoy is that, quite simply, Charles Darwin was the best young man available for the job.

By a happy coincidence, the *Beagle*'s first intended port of call was Tenerife, the very island that Darwin had dreamed of visiting with Henslow. Less happily, however, because of a cholera outbreak in Britain, when they arrived in the harbour of Santa Cruz, on Tenerife, on 6 January 1832, FitzRoy was informed by the authorities that a quarantine of twelve days would be imposed before anyone from the ship would be allowed ashore. Unwilling to delay, he gave orders to up anchor and continue south. So the first place that Darwin actually set foot ashore outside England on the voyage of the *Beagle* was, as we mentioned in Chapter 3, Santiago (also known as St Jago), in the Cape Verde Islands, some three hundred miles off the coast of West Africa and a little over fifteen degrees north of the Equator. They stayed for three weeks while FitzRoy made geographical and magnetic surveys, giving Darwin ample time to explore. While drinking in the profusion of wildlife and the strange geological features of the volcanic island, Darwin made the first significant scientific observation of the voyage. About thirty feet above sea level, there was a white band of material running through the exposed rocks. The white material was made of shells and coral, squeezed and compactified by the weight of the rocks above. But such a band of rock could only have been formed on the sea bed. As Darwin put it: 'A stream of lava formerly flowed over the bed of the sea, formed of triturated recent shells and corals, which it has baked into a hard white rock.'[3]

This could only mean that the sea had once been at least thirty feet higher than it is today, relative to the rocks of Santiago, so that these marine deposits could have been laid down. But was the band of crushed shells and coral so far above sea level now because the sea level had fallen? Or was it possible that the island itself had risen up, out of the sea?

Among the huge library of books carried on the tiny ship (245 volumes, including the *Encyclopaedia Britannica*), Darwin's two most treasured companions were Humboldt's *Travels*, which had been given to him by Henslow, and the first volume of Lyell's *Principles of Geology*. There was a double irony about the Lyell volume. It had been given to Darwin by FitzRoy, who would later be a violent opponent of the idea of biological evolution (an idea which Darwin developed in no small measure under the influence of Lyell's ideas about the age and evolution of the Earth), and although Henslow had advised Darwin to read the book, he also advised him, in a letter, 'on no account to accept the views therein'.[4]

Some of Darwin's contemporaries would have seen the white band as evidence of the Biblical Flood. Others, probably including Sedgwick, who subscribed to the catastrophist school of thought, might have thought in terms of a sudden upheaval lifting the island to its present position. But Lyell wrote of *gradual* changes, with land slowly rising in one part of the world in balance with a slow fall going on somewhere else.

The band of shells and corals, although squeezed by the weight of rocks above and baked by the heat of the lava flow, showed no signs of the kind of massive destruction that should have been caused by a sudden violent uplift of such an extent. Also, the height of the band varied around the island, which Darwin interpreted as a sign that in places there had been some relatively gentle subsidence after the uplift had occurred. At the very beginning of the voyage, at the start of his career as an independent scientist, Darwin was presented with evidence that he saw as supporting Lyell's views – that at least for islands such as Santiago, the landscape had indeed been moulded by gradual changes occurring over long periods of time. This was to become a theme running not only through his studies during the rest of the voyage, but also through his life's work.

As we have seen, such ideas were reinforced at almost every port of call. The fossil remains of South America pointed to gradual change and a long history of the Earth, and the samples of those fossils that Darwin sent back to England did as much as anything to make his name (as a geologist, since fossils came within the jurisdiction of the hammerers) before he even returned home himself. In April 1834, while the *Beagle* was at anchor in the mouth of the Santa Cruz river, on the eastern side of the southern tip of South America, Darwin was part of an expedition up the river and into Patagonia. Here he found more evidence of gradual uplift of the land. The river valley itself was several miles wide, and edged by cliffs about three hundred feet high. On either side, horizontal plains stretched away from the cliff tops. There Darwin found shells and shingle, indicating that the entire flat plain had once been under water.

Gradual uplift could account for that, as well as for the white band on Santiago. But by now Darwin was toying with an even more bold idea. Could gradual uplift account not only for the presence of marine sediments a few thousand feet above sea level, but also for the existence of the entire Andes mountain range? Was *all* of South America land that had been gently lifted up from the bottom of the ocean?

He found more evidence to support the idea of uplift on the western coast of South America a few months later. Near Valparaiso there were seashells in strata 400 metres above sea level, and flat plains lying between the present-day shoreline and the foothills of the Andes, suggesting a former sea floor that had been raised up by geological activity. Soon Darwin was to witness the power of geological activity first hand.

Doubling back southward from Valparaiso, then working their way slowly up the west coast of South America and surveying as they went, on 19 January 1835, the *Beagle* and her crew were at San Carlos, on the island of Chiloé. That night they were presented with a spectacular show by the eruption of the volcano Mount Osorno, some 110 kilometres to the north. In his diary, Darwin described the magnificent sight: 'By the aid of a glass, in the midst of the great red glare of light, dark objects in a constant succession might be seen to be thrown up and fall down.' By the morning, the volcano was quiet, and the *Beagle* stayed at San Carlos for another two weeks, which gave Darwin time for more explorations and to discover, among other things, yet another band of seashells more than a thousand metres above sea level.

Moving northwards again in February 1835, the next opportunity for Darwin to spend some time ashore came at Valdivia, where he explored the local forest. There, on 20 February, he experienced a major earthquake first hand:

> I was on shore and lying down in the wood to rest myself. It came on suddenly and lasted two minutes (but appeared much longer) . . . There was no difficulty in standing upright; but the motion made me giddy. – I can compare it to skating on very thin ice.[5]

The town of Valdivia presented a dramatic sight, with the wooden houses knocked about by the quake; but there was much worse devastation to be seen when the *Beagle* left to continue its voyage up the coast, to the large city of Concepción, more than seventy miles to the north.

They spent ten days surveying the coast up to Talcahuano, Concepción's port, and found it littered with the debris left by a huge tidal wave that had struck following the earthquake. Talcahuano and Concepción were both devastated, and made the deep impression on Darwin described in Chapter 3.

The violence of the earthquake, and the destruction it caused, were alone enough to set Darwin thinking again about the power of geological forces. But at Talcahuano he saw something even more important to his developing ideas. There were fresh mussel beds just above the high-water mark, with the mussels themselves all dead. Before the earthquake, they had been in the tidal zone; now, they were a yard or so out of the reach of the water. The land had been uplifted by a visible amount, by the very earthquake that Darwin had experienced. He had both felt and witnessed the very process which, he was now sure, had raised the Andes themselves to their present great height over an enormous span of geological time.

His major expeditions into the Andes confirmed this growing conviction, as the discovery of fossil fishes far above sea level, petrified forests, and jumbled geological terrain revealing the great forces that had been – and still were – at work in raising the Andes all helped to complete the picture. Powerful but slow-motion forces were clearly at work there, and he wrote about his discoveries and ideas in the letters that he sent back to Henslow.

But this was only half the story. If Lyell was right, the uplifting of the Andes and South America must be counterbalanced by land sinking beneath the sea somewhere else. While Darwin accepted this part of Lyell's theory, however, he began to doubt one detail of Lyell's speculations.

If the Andes were indeed rising, as Darwin now believed, it was natural to think that the floor of the deep Pacific Ocean might be sinking in compensation, producing a kind of see-saw effect. Far out in the Pacific, Darwin knew, lay the coral islands that had been described by other travellers. Some of these really are islands, surrounded by a circular reef of coral; others consist only of the circular coral atoll itself, surrounding a lagoon. The reefs are built up by living creatures, the corals themselves, which need both warmth and sunlight, and live only in shallow tropical water. Many people, including Lyell, believed that the reefs grew around newly formed islands, the rims of volcanic mountains that were rising out of the sea. But Darwin, now firmly his own man and not afraid to disagree even with Lyell where he saw the need, realised that this argument was upside down. He speculated that the coral atolls actually fringed *sinking* underwater mountains, growing upward as the land beneath them sank beneath the waves, until all that was left was the circular fringe of coral surrounding a lagoon where an island had once been.

Happily, part of FitzRoy's brief was to call in at coral islands on the return voyage to England, to take soundings and find out whether the circular reefs really were built on the peaks of submerged volcanoes. Those soundings, and Darwin's own careful observations, confirmed that he was right. In their search for light, the corals build upward as the land beneath them sinks, literally standing on top of the dead coral base built by their ancestors when the sea floor below was nearer to the surface of the ocean.

We now know that Darwin and Lyell were wrong in their suggestion that the entire Pacific/American region was tilting, going down in the west and up in the east. Today, the geological activity that built the Andes is explained in terms of the theory of plate tectonics and continental drift, with South America literally moving westward and riding over the thin crust of the Pacific sea floor, while its 'leading edge' crumples to make the Andes. The Pacific floor itself is also being destroyed at its western edge, pushed under the continental mass of Asia (in compensation, the Atlantic Ocean is getting wider, spreading out from a great north–south crack known as the Mid-Atlantic Ridge, where molten rock from below the crust oozes out and sets as it spreads to form new sea floor crust). But there *is* a downward slope of the Pacific sea floor towards the western rim, and as the oceanic crust is pushed down that slope as if it were part of a great underwater conveyor belt, mountain tops that were once islands do disappear beneath the waves as they grow their fringes of coral reef.

Darwin worked on his geological studies on the way home, turning a mass of notes into something that, he thought, might be suitable for publication. But the way had already been prepared by Henslow, who had been so impressed by Darwin's letters describing his geological findings that he had had ten of them edited and printed up as a booklet for private distribution, before the voyager even returned. Significantly, also, as the ship reached the end of her voyage at the beginning of October 1836 Darwin wrote ahead from Falmouth, their first port of call in England, asking Henslow to propose him for a Fellowship of the Geological Society, not knowing that almost a year earlier, in November 1835, Sedgwick himself had read an account of Darwin's South American discoveries to the Society. It had roused great interest, presenting some of the first evidence that large areas of that continent had been raised over a long period of time by gradual uplift. Darwin's election as a Fellow of the Geological Society (of which Lyell happened to be President at the time) was a formality – and it is some indication of how Darwin viewed his own role in science at the time that although he was in such haste to be elected a Fellow of the Geological Society, he did not bother to join the Zoological Society until 1839.

After visiting family and friends in Shrewsbury, Darwin

reached London at the end of October 1836. On the 29th he had dinner with Lyell himself – the first time that they had met. Lyell, who had seen the basic outlines of his once-controversial theory fleshed out and made respectable by Darwin's discoveries, was more than happy to accept his error about coral islands in exchange for evidence of South America rising up 'at the rate of an inch in a century'.[6] On 4 January 1837 Darwin read his first short paper to the Geological Society, describing this uplift, and by 17 February he had been elected to the Council of the Society, on the strength not only of this work but of his fossil discoveries, which were transforming the understanding of past faunas, and provided support for the ideas of Richard Owen, that fossil species are closely related to the living animals that have replaced them.

Amid all the turmoil of his first full year back in Britain, Darwin picked up on this idea, and it was in July 1837 that he started his first notebook on *The Transformation of Species*. But it would be many years yet before Darwin became known as the father of evolution. The first great project he set himself was to write a technical book about the geology of South America – but that was to turn out a much longer task than he anticipated.

On 7 March 1838 Darwin presented his longest geological work to date, a paper on *Volcanic Phenomena and the Elevation of Mountain Chains* to the Geological Society, arguing the detailed case that the Andes had indeed been raised up by the same processes which were responsible for the volcanic and earthquake activity so common along the western coast of South America. The lively debate that ensued proved to be a turning-point in the battle between the Uniformitarians and the Catastrophists; as Lyell wrote soon afterwards, in a letter to his father-in-law:

I was much struck with the different tone in which my gradual causes was treated by all ... from that which they experienced four years ago [when they were treated] with as much ridicule as was consistent with politeness in my presence.[7]

Hutton had laid the groundwork, and Lyell had initiated serious debate on the subject; but it was Charles Darwin, not Charles Lyell, who convinced the scientific establishment in England that the Earth had indeed been shaped by the more or less uniform action of processes that we see at work today, acting over very long periods of time, and that the continents had been gradually uplifted from the sea, not formed in some great cataclysm.

But at the same time that he was achieving this success, Darwin was growing increasingly uneasy about the damage he might do to his scientific reputation if he told people the way his thoughts about transmutation of species were developing. William Whewell, who was by now President of the Geological Society, came in for particular scorn in Darwin's private notebooks because:

> He says length of days adapted to duration of sleep of man.!!! whole universe so adapted!!! & not man to Planets – instance of arrogance!![8]

But, as Darwin knew only too well, many of his contemporaries saw this nonsense as a piece of profound insight. How would such people react if he made his biological ideas public? It made much more sense for a young man still making his mark in science to stick to geology, where he had already established a solid reputation, than to venture out in public with his embryonic ideas about evolution.

The next landmark in the establishment of that reputation came on 24 January 1839, just over two years after the end of the *Beagle* voyage, when Darwin was elected a Fellow of the Royal Society, still a couple of weeks short of his 30th birthday, and a few days before his wedding. By now, the grand scheme to write a book about the geology of South America had been temporarily put to one side, and Darwin was concentrating his efforts on a book about coral reefs. He had long since completed his account of the voyage of the *Beagle* itself, but the publication of his *Journal* had to wait until FitzRoy completed his own account, so that the two reports could be published together.

When Darwin had been working on his *Journal*, in November 1837, he had written to Henslow to express his feelings on the efforts he had put in to it:

If I live until I am eighty years old I shall not cease to marvel at finding myself an author: in the summer, before I started, if anyone had told me I should have been an angel by this time, I should have thought it an equal improbability. This marvellous transformation is all owing to you.[9]

It was a year and a half after that letter was written, and two and a half years after the *Beagle* had returned to England, when the *Journal* was first published, in May 1839, under the rather intimidating title *Journal of Researches into the Geology and Natural History of the Various Countries Visited by* H.M.S. 'Beagle,' *under the Command of Captain FitzRoy, R. N., from 1832 to 1836.* But what is not immediately apparent (at least to a modern eye) from the title was very much apparent in the text itself – Darwin was not just a great scientist, but also an excellent writer who communicated his discoveries and ideas clearly and entertainingly. Copies of the *Journal* were sent out to friends and colleagues, who responded in glowing terms with praise for the work. Von Humboldt himself, one of Darwin's heroes, joined in the chorus. And, in a sure indication of the clarity of Darwin's prose, even those few reviewers who did not like his conclusions had undoubtedly got the message he was putting across – one writer objected that if Darwin's ideas about the slow rise of South America were correct, then 'at least one million of years must have elapsed' since 'the sea washed the feet of the Cordillera of the Andes!'[10]

The comment was intended to ridicule Darwin, but this response was not typical. Indeed, the public response to Darwin's *Journal* was so great that by August it had been published separately, to the chagrin of FitzRoy, who now argued publicly that all of the evidence that Darwin interpreted as indicating the uplift of the Andes was a direct result of the Biblical Flood.

Darwin's work on his book about coral reefs continued throughout the rest of the time that he and Emma (and their growing family) lived in London, interrupted by long bouts of illness, as well as by Darwin's work for the various learned societies that he was involved with at the time, by zoological investigations including a study of fishes, and, of course, by his continuing secret speculations about the nature of evolution. The book was eventually published in 1842, more than three years after Darwin began work on it, but this was not another *Journal* to make the public sit up and take notice of him. Instead, it was a technical work, which as Darwin himself wrote was 'thought highly of by scientific men'.[11] This book established once and for all the explanation of coral atolls in terms of islands sinking slowly beneath the waves, an explanation that still stands, with only minor modifications, today.

So by the time the Darwins moved to Down House, he was less of a public figure than he had been in the heady days after the return of the *Beagle* and at the time of the publication of the *Journal*, but a solidly established scientist with an excellent reputation for his work in geology. That reputation was consolidated in the 1840s. He wrote a book about volcanic islands, published in 1844, and eventually completed his *Geological Observations on South America*, delayed by the interruptions caused by his illnesses and by his other interests, in time for it to be published in 1846, ten years after the completion of his voyage aboard the *Beagle*. It was the last of his publications about the observations made on the voyage itself, and his relief at completing a long and sometimes tedious task shines through in a letter he wrote to Henslow in October 1846:

> You cannot think how delighted I feel at having finished all my *Beagle* materials except some invertebrata: it is now ten years since my return, and your words, which I thought preposterous, are come true, that it would take twice the number of years to describe, that it took to collect and observe.[12]

But before moving on to the next phase in Darwin's life, there is one more aspect of his geological work which is worthy of mention – not so much because of its relevance to geology, but because of the insight it provides into Darwin himself, as a scientific thinker.

In June 1837 Darwin had visited the so-called 'parallel roads' of Glen Roy, in the Scottish Highlands. These are three ledges on either side of the narrow glen, each horizontal and at the same height as its counterpart on the other side of the valley. After a week of investigations Darwin decided that the ledges were beaches that had been formed when the glen had been filled by an arm of the sea, and he wrote a paper to that effect which he published in 1839.

Over the next two decades the interpretation of the 'roads' remained a subject of minor, but lively, debate among the geological community. This was just at the time when the theory of Ice Ages (or at least, an Ice Age) was being strongly promoted by the Swiss scientist Louis Agassiz (1807–73) and his followers, and it seemed more likely, to some geologists, that the ledges in the glen had been formed by the action of a lake associated with glaciers than by sea water filling the valley.

Darwin himself was receptive to the idea of an Ice Age – indeed, in 1841 he published a paper describing the way so-called 'erratic' boulders had been transported by ice flows from their original locations to their present resting places in South America, dropping them among geological features where they do not belong. This was one of the first Ice Age papers by an English scientist, but in spite of this pioneering contribution Darwin remained convinced that sea water, not a glacial lake, had carved the ledges in the sides of Glen Roy.

In 1861, Darwin suggested that the geologist Thomas Jamieson should visit Glen Roy to resolve the issue. The investigation carried out by Jamieson showed that Darwin was wrong – that the ledges in the sides of the glen were indeed the shorelines of a former glacier lake. And Darwin immediately retracted his idea in the face of solid scientific evidence. He wrote to Jamieson that:

Your arguments seem to me conclusive ... I give up the ghost. My paper is one long gigantic blunder ... I have been for years anxious to know what was the truth, & now I shall rest contented, though ashamed of myself.[13]

This, of course, is the way scientists are supposed to behave, modifying their ideas (and even abandoning them altogether if need be) in the face of observational or experimental evidence that they need to be modified. But such an attitude is not always the way scientists respond to evidence that their ideas are flawed. Like other human beings, many scientists like to cling on to their cherished beliefs for as long as possible, and often find it hard to accept, let alone acknowledge publicly, the error of their ways when they are wrong. But Darwin rose above all that, as great scientists always have done, in his quest for truth. What mattered to him was indeed the truth, not whether the evidence supported his own pet hypothesis.

That search for, and love of, the truth led him to the theory of evolution, no matter how uncomfortable he must have felt about the inevitable conflict with the religious establishment in general, and with the views of Emma in particular. But over the eight years following the publication of his *Geology of South America*, while the evolutionary speculations continued in private, Darwin's main public scientific preoccupation would be with what had seemed in 1846 to be a trivial footnote to the story of the *Beagle* voyage, those invertebrata that he mentioned in his letter to Henslow. Barnacles became the theme (and sometimes the bane) of his scientific life, as Darwin the geologist evolved into Darwin the biologist, surrounded by an ever-increasing family and tended by a loving wife in the calm of the Kentish countryside.

Notes

1, 4, 5, 7, 11, 12 John Bowlby, *Charles Darwin*, Hutchinson, London, and Norton, New York, 1990.

2, 3 *Autobiography*.

6, 8 Quoted by Adrian Desmond and James Moore, *Darwin*, Michael Joseph, London, 1991.

9, 13 Letter to Henslow quoted by Ronald Clark, *The Survival of Charles Darwin*, Weidenfeld & Nicolson, London, 1985.

10 *Athenaeum*, (15 June 1839), no. 607, p. 446.

Chapter 7

Seclusion

The house in which Darwin spent the last 40 years of his life stands today almost untouched since the end of the nineteenth century. On the surface everything appears to be the same – the lush Kent countryside, the manor house at the edge of the village of Downe* the stillness and seeming isolation from the metropolis a mere 20 miles away. But, Down House is a gentleman's country home preserved in aspic, a ghost. In the 1840s and 1850s it was a rambling family home dominated by the growing Darwin clan, children's voices, hushed as they passed their father's study on the ground floor, servants taking care of every domestic need, eminent scientists visiting from London, and in the early days at least, the hammering and banging of workmen making improvements to the house.

The family arrived at Down House on 17 September 1842. Although it was only two hours' carriage ride from the centre of London, Darwin felt that by moving there he had finally achieved his ambition of escaping to the country, taking Henslow as his paradigm – the idyll he had dreamed of as he was tossed around on the high seas south of Cape Horn.

Although the house was large and ideally situated, it was in rather bad decorative order, and with the family growing it was

---

* The spelling of the village name changed to Downe in the middle of the 19th century, but the original spelling of Down House was kept.

obvious that some extensive restructuring and redevelopment would have to be undertaken to accommodate them. The room Charles had ear-marked as a study was eighteen feet square and lined with bookshelves to house his library and papers. Two tall windows along one wall brightened the rather drearily papered walls, but even this substantial room had to be extended later as the library and Darwin's correspondence grew.

That first autumn the decision was made not to get started straight away on redecorating because Emma was almost nine months pregnant and they both agreed that the baby should be delivered before the builders set to work. Little more than a week after moving in, Emma gave birth to their third child, Mary Eleanor. Born tiny and weak, she lived only three weeks.

Emma was naturally devastated at the loss and for a while a gloom settled over Down House – a poor start to their new lives and eased only by the fact that William, now nearly three, and eighteen-month-old Annie appeared to be in good health. The Darwins' approach to overcoming this and other tragedies befalling them was to get on with life stoically and within three months Emma was again pregnant.

Their first winter at Down convinced Darwin that in the spring the builders would have to be brought in and the house altered. There were draughts everywhere, the exposed north side of the house took a battering from severe gales and the snow piled up against the stone walls. With Dr Robert's financial assistance – £300 provided against Charles' inheritance – the work was started as soon as the weather improved. During the course of the next six months the house was transformed. A new bedroom and a schoolroom were built, the existing bedrooms radically altered and the kitchen and pantry enlarged. A new high wall was built around the property, an orchard planted in the grounds and, with the support of the local authorities, a road which passed directly beneath the study window was lowered three feet and screened by a wall. By the summer most of the work had been completed and the rooms decorated.

In some ways the first decade the Darwins spent at Down House encapsulated for them the best and the worst of times. It was a period during which Emma gave birth a further seven times, a decade during which Charles found further evidence to support his theory of evolution and made at least one close friendship with a man who would be of enormous help to him and would remain a life-long confidant. It was an era preceding the stresses and turmoil of the public exposure of Darwin's theory and a time of seclusion from political unrest and the dirt and grime of the big city. On the other side of the coin, the second half of this decade, from early 1847 to 1851, marks the period when Darwin experienced some of his worst health problems and the family suffered loss and tragedy.

Darwin first came into contact with his life-long friend Joseph Hooker through the latter's participation in an Antarctic expedition captained by Sir James Ross. Hooker, who had been trained as a doctor in Glasgow and was the son of Sir William Hooker (whom he succeeded as Director of the Royal Botanic Gardens at Kew), was himself an expert botanist and had spent the last four years investigating the plant life of the southern hemisphere. Hearing of Hooker's work through Sir William, Darwin was immediately interested in off-loading his own specimens collected from the voyage of the *Beagle*, which had still not been studied. Within weeks of Hooker's return, Darwin had written to the younger man offering him his entire collection of plants from the *Beagle*. Realising the specimens would be a perfect supplement for his own collection and wanting to establish personal contact with Darwin, whose own reputation was in a steep ascendant, Hooker naturally jumped at the chance.

The two men had much in common. They had both studied medicine and then embarked on a scientific expedition at almost the same age; they also shared many social and political views and had similar tastes and interests. Darwin was the older of the two by eight years, and Hooker initially perceived him as an established expert but a man close enough to his own age and social standing to be accessible. Soon their relationship developed from being a meeting of minds on scientific subjects to intimate friendship.

This was an important relationship for Darwin. Although he had a number of friends and a wide-ranging collection of colleagues, he was not, with the exception of his beloved wife, really close to anyone near his own age. With Emma he shared most things; they were intimates and their relationship meshed even more as their family grew around them. Emma was Charles' anchor, she mollycoddled him through his many and various illnesses and acted as his emotional crutch when work overwhelmed him. They leaned on each other during the many times of tragedy and stress throughout their long marriage and were a devoted couple. But, despite this intimacy, Charles could not discuss science with his wife and, for her part, Emma could never accept his lack of religious faith. Although she read his work and even gave him moral support despite the conflict she experienced between Charles' ideas and her religious faith, she could never be a colleague, and perhaps for the sake of their marriage it was better that way.

Then there was Eras, whom Charles loved and was close to. He could discuss and argue science with Charles and was independently interested, but he was not a professional scientist and was far more concerned with having a good time than participating in the world of discovery of which Charles was so enamoured. In Hooker, Charles found a man who was a combination of professional companion and personal friend, a man in whom he could confide his most secret, radical ideas. It is a sign of how quickly and readily the two men became intimate friends, and at the same time just how desperate Charles was for a scientific twin, that within two months of his first correspondence with Hooker he told him the vague outline of his ideas on evolution.

At first his approach via letter was timid, almost self-deprecating. Using expressions such as 'engaged in very presumptuous work, perhaps a very foolish one' and 'after blindly collecting', he finally came out and stated the unpalatable truth at the core of his ideas: 'I am convinced (quite contrary to the opinion I started with) that species are not (it is like confessing to a murder) immutable.'[1]

As can be seen from the way Darwin compares the effort with the confession of a murder, his admission was of course an astonishing claim, and although it was not received with the same chill as Lyell's response to the same notion two years earlier, Hooker did not immediately agree. In fact, although later

Hooker became one of Darwin's most important and consistent supporters, at the time Darwin first mentioned his ideas, he was far from convinced. It was to be many years, and after he had collected his own observations and worked through the theory in his own way, before Hooker came round to accepting the validity of his friend's claim. Once convinced, he was unerring in his support.

During the course of the next three years Darwin and Hooker grew closer both personally and intellectually. Hooker made several visits to Down, sometimes staying for two or three weeks at a time, during the course of which he became a good friend to Emma as well as Charles. When he was unable to visit, the two men corresponded weekly. In the course of their letters they discussed everything from domestic matters and Charles' continuing symptoms of skin complaints and sickness to the highest intellectual planes of Darwin's theory of evolution and matters pertaining to their plant collections from South America. When they discussed evolution, Darwin found Hooker's independent stance a great help. In some ways, through obvious necessity, Darwin had become too isolated, too inward-thinking on the subject; the freshness of a colleague with whom he could discuss in cold scientific terms the most outrageous heresies and socially explosive notions was a tremendous help, even if – or perhaps because – the other half of the partnership was an ardent sceptic.

Through these talks with Hooker it became obvious to Darwin that he must put his theory into better material shape. While living in Upper Gower Street he had collected voluminous notes on the theory and had discussed breeding theories with farmers, read everything available on Lamarckism and the theories of other evolutionists. He had matched up his heretical ideas with his observations taken during the voyage of the *Beagle* and made comparisons with Sedgwick's often radical geological theories. He now felt ready to put the theory into some sort of tangible, readable form.

The first version of what would fifteen years later see publication as the *Origin of Species* was begun during the spring of 1844 and was originally little more than a sketch. By the summer it had grown into a manuscript 189 pages in length and after it had been transcribed as a fair copy by the local schoolmaster, it had grown to 231 pages.

Thanks in part to his ill-health, already by his mid-thirties Darwin was becoming increasingly aware of his own mortality. For some time he had been privately fretting about the state of his theory and how in the event of his premature death, his ideas would be lost for ever. Now, with his work written up, even in relatively unpolished form, he felt a great sense of relief. What was more, he could let those closest to him read the theory for themselves.

First in line was not Hooker but Emma. Charles knew that the central theme of the theory, which they had discussed many times, would be anathema to her but he felt that she had to see it first. He had no intention of submitting it for publication, the time was clearly still inappropriate; but he had Hooker in mind as someone whose professional opinion he valued and whose views he would want to gauge at the correct moment. Certainly, and rightly, Emma would never have forgiven him if he had by-passed her.

Emma's opinion of the work was a delicate balance of tact and disapproval of the central tenets. She suggested improvements to Charles' prose and queried a few points. Otherwise she was proud of her husband's effort but non-committal about the main thrust of the subject, which was completely at odds with her own orthodox Christianity. But, whatever Darwin felt about Emma's response, he continued to keep the essay under his hat for some time and did not even show it to Hooker until January 1847, a full two and a half years later.

During this time Darwin's research and writings went through a number of changes. He started work on another book about his travels, *Geological Observations on South America*, which was published in 1846; later that year he started studying the barnacle, a task which occupied almost all of his time for the next eight years, details of which are recounted in the next chapter. It was also a period during which Darwin began to adopt the mantle of the country squire.

In his mid-thirties with a growing family to take care of and his scientific reputation still in the ascendant, he decided after consultations with his business-minded father that the time was right for him to invest in land. Guided by his father's expert knowledge of land and property development and after appointing an agent to search for the best bargain, during the spring of 1845 Darwin acquired 324 acres of land in Lincolnshire. It cost £12,500 and included a farm and a number of dwellings which were rented to the incumbent tenants. It was calculated that the land would return a handsome annual sum, at least £400 from the farm alone, and provide the family with long-term security. It also brought with it a squire's responsibilities, for which, initially at least, Charles was not fully prepared.

Although it was considered by financial advisers to be an auspicious time to invest in land, and although the possession of real estate then, as during almost any era, was seen as a protection against most of the political and social ravages of the day, through a freak event no financial expert could have predicted, it actually turned out to be bad timing. During Darwin's first year as a landlord, in the summer of 1845, the potato famine struck.

The famine came about as the result of a fungal infection of that and the subsequent year's potato crop, a blight which developed into one of the most devastating natural disasters to hit Europe. During the following five years the famine took the lives of 700,000 in Ireland alone and prompted the emigration of over a million people. Although not as badly hit as Ireland, the crop in England and Wales was destroyed and starvation took the lives of thousands of peasant farmers, labourers and their families.

At first the famine did not affect the Darwins, but when the locals began losing their crops to the disease and found themselves with only a few weeks of potato stocks laid in, the family became alarmed. At the time a ridiculously high import tax on corn meant that bread was prohibitively expensive, making the loss of the potato crop far more serious. Under protest, the government did eventually see sense and cut the tax so that bread became more affordable, but the change in the law came too late to save thousands of lives. Darwin and many other Whig landlords took the subsequent drop in the value of their farmland in as egalitarian a manner as possible. Through his agent, Darwin allowed his rents to drop by fifteen per cent to help compensate for the farmer's decrease in income. Emma handed out penny bread-tokens to the local workers, some of whom were back in Down House that summer making still more alterations, and after the foreman working on the house sacked the entire workforce because they were bickering instead of working, Charles insisted that they be reinstated.

The repeal of the Corn Laws and the lowering of rents helped some, but the majority were beyond such measures and the partial loss of the potato crop for the next two years further exacerbated the problem. Robert Peel's Conservative Party, which had been born from the ashes of the Tories (devastated a decade earlier by the First Reform Bill) and which had toppled the Whigs at the last election in 1841, were in their turn ousted in 1846 by the furore over the Corn Laws. Peel saw the wisdom of repealing the outmoded laws and revived the income tax system, but it destroyed his career. In his last great act of political power he pushed through the change in the law but lost his position by an unseemly temporary alliance between a group of high Tories and the Whigs.

Although governments fell, democracy was preserved and despite the turmoil of the times, the continuing efforts of the Chartists to destabilise society got them nowhere. Peel had lost his job because of his determination to repeal the Corn Laws, a sacrifice which Darwin, along with the majority of the general public, greatly admired, but the Whig government that came to power as a result and remained there for the next 20 years pleased him even more. This radical shake-up in British politics, combined with a stabilisation of the country's finances created out of Peel's five-year term as First Lord of the Treasury, ushered in a period of relative political calm.

Throughout 1846 and 1847 the friendship between Hooker and Darwin grew steadily and Darwin was placing more and more trust in the younger man. They had their arguments and in May 1847 a more serious falling-out over a scientific issue – the geological origin of coal, which lead Darwin to say of his colleague: 'There is no man more lovable and none more "peppery".'[2] That said, at the same time, he often told colleagues that he had extracted more from his friendship with Hooker than from any other person.

Meanwhile, by late summer 1847 Darwin was again falling prey to a new succession of illnesses. Although he had never been entirely free from nausea and skin complaints, stomach ailments and headaches for any length of time, retrospectively, the period between moving to Down in September 1842 and the summer of 1847 was one during which he enjoyed relatively good health. Like almost every facet of Darwin's health, the cause of this new bout of illness is a question of debate.

We discussed in Chapter 5 the theory that Darwin's health problems were caused by multiple allergic response aggravated by stress, and it is interesting to note that during the second half of the 1840s his symptoms showed a worsening at precisely the point in his life when a number of domestic problems beset the family and caused Darwin inevitable stress. It is possible that, taken cumulatively and building up during his relatively healthy period between 1842 and 1847, these stressful events could have greatly affected his health.

First, in June 1846, Emma's mother, Elizabeth died and Emma had to visit Maer for some time during the summer to help arrange the funeral and to assist her older, spinster sister Elizabeth, now the sole member of the Wedgwood clan left at the family home (Josiah Wedgwood had died in July 1843). During her absence Charles fretted endlessly and in daily letters he complained about his many symptoms and his loneliness. Yet he had plenty to do. Working as he was on *South America*, corresponding with his land agent in his new role as a country squire and continuing to improve the manuscript for his theory of evolution meant that, in the short term at least, he had more than enough on his plate. What really pained him was the fact that he had to rest so often between short spells of work. Alone on the sofa or looking out across the fields, he yearned for Emma and poured his heart out in long, emotional letters to her.

It was not all doom and gloom. He enjoyed visits from Hooker and other scientific colleagues and when he felt well enough he made the occasional foray into London where he attended some of the monthly council meetings of the Geological Society and, less frequently, made visits to the Royal Society. Although he loved the solitude Down provided, he still valued interacting with his contemporaries and his friends within the scientific community and wrote regularly to Owen, Lyell, Henslow and others.

Two years later it was Charles' turn to attend a gravely ill parent, when, during the spring and summer of 1848, he was obliged to leave the family at Down house and make frequent and lengthy visits to The Mount, where his father was seriously ill and close to death.

Charles hated being away from his family but needed to be by his father's side. Robert had been Charles' cornerstone and one of the most influential people in his life. At The Mount Charles stayed in his old room surrounded by memories of his childhood and teenage years, but it was a very different place now. The house had been a miserable place ever since Charles' mother had died 31 years earlier and Robert had fallen into permanent grieving. Yet, despite this, Charles' childhood had been relatively carefree. In those days he had had little notion of what he was to do with his life. In some respects the chances of early promise had been foreshortened by privilege which had led to a directionless existence until his father had taken a strong hand in pushing him into medicine. The Mount may have been dreary but as a child Charles could always escape to his and Eras' lab or ride off across the estate to Shrewsbury, go hunting or fishing; even during winter or when the weather was unseasonable he could lose himself in the contents of his father's vast library. Later there had been the attractions of Fanny Owen, shooting parties and dances at Maer. Now, everything was different. Fanny Owen was unhappily married, poor Frances Wedgwood had died young and here he was a married man with a growing family to take care of, responsibilities as a landowner, and weighed down with the burden of his heretical work. But, above all else was the fact of his ill-health. In those long, golden summers and throughout his youth he had been perfectly healthy. He had ventured on a five-year trip around the world, an experience almost unheard of during the nineteenth century. Yet, with the singular exception of his illness at Valparaiso, he had suffered little more than a cold during the entire voyage. Now, here he was still only in his late thirties and suffering from a catalogue of symptoms which never seemed to subside. For much of the time he spent at The Mount he was in bed or laid up on a sofa in the library almost as much an invalid as his 82-year-old father.

During Charles' first visit during the spring of 1848, Robert told him that he would suffer a protracted illness and then die suddenly. He must have been a very good, intuitive prognostician, because that is exactly what happened. There were times during which he fell into semi-consciousness broken by spells of unusual lucidity and even one or two occasions when he was able to leave his bed and sit wrapped up in blankets in the garden or the library. On his final visit in October, Charles stayed two weeks and left the old man looking unusually cheerful and comfortable. That was his final memory of his father because on 13 November he heard news that Dr Robert had died peacefully in his sleep.

Charles was undoubtedly devastated by the news; his father had always been his crutch, a man who had guided his son as much as he could and was a powerful, domineering personality. Although Charles had gone largely his own way in the end, he would have been nothing without his father. If Charles Darwin had not been born into a wealthy family and had not been given the active intellectual support and guidance of Dr Robert, he may never have succeeded academically. Without his father's eventual approval and financial support he could not have accepted FitzRoy's invitation to go on the voyage of the *Beagle* and without that, he could never have developed his world-altering theory of evolution. Part of the problem lay in the fact that Robert Darwin's death, although drawn out and in the end inevitable, came at a particularly bad time for Charles. During the mid-1840s he had grown close to Joseph Hooker and had begun to rely on his friend for intellectual support during the development of his theory. Hooker had become a confidant and safety valve for Darwin. Then, in April 1847, not long after he had been given a copy of the evolution essay, Hooker dropped the bombshell that he was embarking on another expedition.

Ever since returning from the Antarctic expedition in 1843, Hooker had been gripped by wanderlust and was desperate to escape abroad again. In the meantime he had fallen in love and become engaged to Henslow's eldest daughter Frances, but even this attachment could not stop him from accepting a place in a team travelling through the Himalayas.

Charles was depressed by the news not least because he was now relying on his friend for help with the fine details of the theory of evolution and his critical opinion of his ideas. Hooker assured him that they could still correspond and even offered the sweetener that he hoped he might find his own confirmation of some of Darwin's ideas during his travels, which might help to convince him of the validity of the theory. It was hardly compensation. Charles needed Hooker far more than Hooker needed him. What was more, he had deliberately chosen to cut himself off from the world; he had few friends and saw his wife and children, Hooker, his brother Eras and their dying father as perhaps his only loved ones. If Hooker was to go he would lose a friend and a helper in one fell swoop. He tried his best to persuade him of the folly and the dangers of the proposed trip but nonetheless found himself travelling to Kew on 20 August to make his fond farewells and to wish his friend luck. It was to be three and a half years before the two would meet again.

Darwin did his best to reconcile himself to the loss and made a conscious effort to distract himself with work. The problem was, he was ill so much of the time that he could hardly manage to work more than three or four hours a day at best and often went for weeks at a stretch unable to overcome the nausea, headaches and feeling of lethargy. He was also engaged in a long-drawn-out and what he suspected would be a thankless task of investigating his barnacle specimens. It was tedious in the extreme and required using a microscope for hours at a time which did nothing for his headaches.

Although Darwin suffered his fair share of domestic and personal problems during this period, ensconced as he was in his idyllic country retreat, the second half of the 1840s was not all bleak. Down was not just a study occupied by a work-obsessed and semi-invalid scientist, it was also home to a growing clan of children. During the course of their long marriage, Emma bore ten children, nine of whom were born by 1851, within twelve years of their marriage; their last child, Charles, born to an ageing Emma at the end of 1856, died aged eighteen months.

The many Darwin children, as well as the children of visiting relatives who sometimes stayed for weeks at a time during the summer, kept Down House alive. Even at his most maudlin, when work was impossible and he was racked by stomach pains and skin boils, Charles was buoyed up by the children whom he adored and with whom he spent as much time as possible. Darwin was very much a family man, a classic Victorian patriarch but without the caricatured cliché-ridden sternness of the unapproachable father figure ruling the home with a rod of iron. He also had a lively interest in many areas outside science. Until middle age he was fond of music and poetry. He and Emma read novels with a voracious appetite. Knowing just how much effort went into a good book, he always had the deepest admiration for writers of fiction. Into middle age he loved Shakespeare and read Wordsworth, Coleridge and Shelley, and as an old man he tried in vain to reread his favourites, but found the spark had somehow gone out of them. In his autobiographical writings he admitted that in later life he found Shakespeare so intolerably dull that it nauseated him and that he could no longer endure reading a single line of poetry.

This admission is quite peculiar and deserves closer analysis. Writing towards the end of his life, Darwin says in his notes about himself that for perhaps twenty or thirty years he had been unable to enjoy the greats of literature he had once adored, which places his disaffection somewhere around the 1850s. What changed his feelings? Why the loss of interest in what he calls 'the higher aesthetic tastes'?[3]

As we shall see later in this chapter, Charles experienced the greatest tragedy of his life very early in the 1850s and this may have affected him emotionally in such a way that he lost interest in poetry, art and highbrow literature, in much the same way that personal tragedy can strip a religious person of their faith in a benevolent God. Alternatively, Darwin may have lost faith in human creativity as a result of his discovery of the mechanism of evolution. By a convoluted and untraceable route it could be that his understanding of evolution somehow diminished his respect for humankind and its creations. Instead, he found he could turn to art only as a means of entertainment, relishing mundane literature for its escapism value, unable to delve into the deeper, more emotional waters of high art because he was able to see through the conceit and vanity it represented.

Whatever the root causes of his disillusionment with art during his forties, the years leading up to this were undoubtedly unsettling. His health would improve for a time, such as the first years at Down, and then be followed by stressful events precipitating a new bout of illness. For eighteen months, from the departure of Hooker in 1847, Darwin slid into a steady decline. He became severely depressed, his work on barnacles ground almost to a complete halt, visitors stopped coming to the house and fresh invitations to family and friends were sent rarely.

Darwin retreated into a private world, alone most of the day and sharing his unhappiness only with his beloved Emma, who grew more and more concerned for him as each day passed.

The worst time followed Dr Robert's death. Already in decline some months before his father had taken to his bed, Charles was deeply affected by his father's passing. He had no God, now he had lost his father too.

Charles' friends became worried about him. By the spring of 1849, he had not been seen by colleagues in London for many months and the rare visitor to Down, such as Darwin's old friend from the *Beagle*, Lieutenant (now Captain) Sulivan, who dropped by before leaving with his family for the Falklands, found him looking grey and drawn, hardly able to walk. Clearly, something had to be done and before leaving, and Captain Sulivan provided Darwin with advice which would greatly improve his health for the rest of his life – he recommended Dr James Gully's Water Cure Establishment in Malvern.

Dr Gully's had been open since 1842 and by the time Darwin attended the establishment it had become a fashionable centre for the wealthy and famous of the day. At first Darwin was highly suspicious of the whole idea and viewed such places with cold scientific detachment and one wary eye on the vast sums of money Dr Gully was reputedly earning from his illustrious clientèle. But gradually, over the following few months, his opinions changed. He talked to friends and colleagues about the water cure and discussed the option with his own doctor, who found he could not advise him either way. Eventually with

Emma's persuasion he came to the conclusion that his health had now become so bad that he should try almost anything if there was any chance at all that it would help him regain a semblance of his former fitness. The snag was that for Darwin to take the treatment he had to spend some time in Malvern. So the entire family decamped, a house was rented in the town not far from the doctor's establishment and Darwin decided to submit himself wholeheartedly to the regime.

The treatment involved cold showers, rough scrubs with hot water and cold compresses worn all day and redampened every two hours. He had to eat a special diet, free of sugar, salt, bacon, alcohol or any other stimulant, and he was given homeopathic medicines. Gully diagnosed Darwin as suffering from a nervous disorder precipitated by too much mental exertion and insisted that Darwin should take a lengthy break from working in his lab, have a long rest and relax. This, combined with a tough regime of early morning walks and water treatment at severe extremes of temperature would, he assured Darwin, effect a cure.

Although Darwin was never convinced of the reasoning behind the treatment, he became a close friend of Dr Gully's and placed his treatment entirely in the doctor's hands. Surprisingly perhaps, the technique worked. Within weeks of his arrival in Malvern Darwin was already feeling better; his heart palpitations had stopped, his headaches were subdued and he no longer suffered from regular and recurring nausea. In fact the treatment was going so well that the family decided to stay in Malvern far longer than they had originally planned. The expected three-week visit was extended to almost four months, the entire Darwin clan arriving back at Down at the end of June 1849.

Despite feeling refreshed and revitalised, Darwin only eased himself back to work gradually. Obeying the doctor's advice to the letter, he had constructed a special bathhouse in the grounds of Down which enabled him to continue with the regime he had started so successfully in Malvern. A carefully constructed cistern delivered a freezing cold shower at the tug of a lever and for many years after first visiting Dr Gully, Darwin stuck to the therapy religiously – a trip to the bathhouse at 7 a.m. each morning, where he had a form of sauna which involved him wrapping himself in blankets and sitting on a chair which was heated by a spirit lamp until the sweat dripped off him, followed immediately by the freezing shower or a plunge in an ice-cold bath.

He was allowed to work only two to three hours a day at first, but even this was better than the two or three hours he managed feeling nauseous and depressed before the Malvern visit. For over a year Darwin experienced the best health he had known since his bachelor days immediately after his return to England. His researches on barnacles moved onward rapidly; he continued improving the text of his evolutionary work and he kept up his regular correspondence with Hooker, who was now beginning the third year of his travels. Then tragedy struck. The most painful experience of Darwin's life and an event which would for ever seal his feelings about God, man and the scheme of life began to unfold in June 1850, when his favourite daughter Annie fell ill.

For some time Darwin had harboured fears that his illness would prove to be of an hereditary nature and he had been watching anxiously for the first appearance of symptoms in his offspring. Until then none had appeared. The children had all suffered from youthful complaints as well as the inevitable colds and flu, but none had shown any signs of the recurring and varied symptoms experienced by their father. But during that summer, nine-year-old Annie went down with a succession of illnesses ranging from stomach disorders to headaches; she would frequently burst into tears for no apparent reason and fell into deep depressions completely out of keeping with her normally cheerful character. The symptoms kept up for the next few months, followed by a short spell of good health when for a week or two she made a marked improvement; but, to Darwin's horror, more symptoms soon appeared – nausea, vomiting, skin complaints, mouth ulcers, cramps and lethargy. Mortified, he could see a reflection of himself in his daughter, a future plagued by a series of incurable, but non-life-threatening illnesses that would make her weak and miserable. It was a dreadful fate he could not wish upon anyone, let alone his favourite child.

After a bad winter, during which Annie was in bed more often than she was out playing, it was decided finally that she should follow her father and take Dr Gully's cure in Malvern. Charles took her there on 24 March 1851 and stayed a few days before leaving her in Dr Gully's trusted care. Two weeks later, after an initial period during which the treatment appeared to be working well, Annie had a relapse and was put to bed. An urgent message was sent to Down and Charles dashed to his daughter's bedside.

Arriving there two days later, he was appalled by the condition in which he found his daughter and hardly able to believe the change which had overcome her in so short a time. When he had left her two weeks earlier she had looked tired and pained but now the girl was a mere shadow of her former self. Slipping in and out of a delirium, she had moments when she could hardly recognise her father and mumbled gibberish. She was feverish and clammy to the touch. Unable to keep her food down, Annie had grown terribly thin, almost to the point of emaciation, and she was deathly pale.

One hundred and fifty years on, it is almost impossible to say for certain what was wrong with Annie Darwin, but from the strange muddle of symptoms and the rapid deterioration she experienced, it appears likely that she suffered an immuno-system collapse. What initiated this and why it should have struck her at this age is impossible to say with any certainty. Her death was slow and horribly painful. She clung on for a further week after her father's arrival in Malvern, a week during which Charles hardly left her side. Emma, who had to stay at home in Down with the rest of the family, was sent daily reports but she had to suffer the agonies of not knowing what was happening for at least 24 hours at a time and received the news without being able to share her sorrow with her husband. Because they could not have each other to comfort, the pain for Emma and Charles was redoubled. Eventually, death came almost as a relief. Annie had suffered so much and towards the end it had become quite clear that she would not recover from the terrible state into which she had degenerated. Darwin was inconsolable.

Despite his misery, at first he insisted that he stay in Malvern for several days to supervise the funeral arrangements, but Emma's sister, Fanny Wedgwood, who had been with Charles and had taken turns with him in nursing the sick child during the past week, insisted that he return home to Down where Emma, who was then eight months pregnant, needed him desperately. Reassured that she could take care of the arrangements in Malvern, Charles knew she was right and submitted to her logic.

Returning through the English countryside on the long trek back to Kent, Darwin felt utterly wretched and would never again experience such deep and agonising sadness. In losing his beautiful daughter – the little girl who had meant so much to him with her perfect character, so charming and gentle, a child who had never knowingly upset anyone and who was bright and intelligent, funny and affectionate – he had also lost any remaining vestige of religious faith he may have had. From that moment on, Darwin was a total, uncompromising atheist: his only god was rationality, his only saviour, logic and science; to that end he would continue to dedicate his life. There was no meaning to existence other than a culmination of biological events. Life was selfish and cruel, headless and heartless. Beyond biology there was nothing.

Notes

1 *The Correspondence of Charles Darwin*, 8 vols, ed. F. Burkhardt and S. Smith, Cambridge University Press, 1985–93, Vol. 3, p. 2
2 *The Autobiography of Charles Darwin*, ed. Nora Barlow, Collins, London, 1958, p. 105.
3 *The Autobiography of Charles Darwin and Selected Letters*, ed. Francis Darwin, Dover, 1958, p. 54.

Chapter 8

Barnacles and Biology

When Darwin had completed the geological work resulting from his voyage on the *Beagle*, he turned his attention, late in 1846, to some peculiar barnacles that he had picked up on the shores of southern Chile in 1835. These strange, tiny creatures had first been mentioned in one of Darwin's letters to Henslow from the *Beagle*, in which he referred to 'a genus in the family of Balanidae' which 'lives in minute cavities in the shells of Concholepas', a kind of mollusc.[1] Dissecting and studying these extremely small creatures (each one only the size of a pin's head) in order to find out where they belonged in the classification of life on Earth would be painstaking work, but at first Darwin delighted in the prospect of getting to grips with what he referred to as his 'beloved barnacles', returning enthusiastically to the kind of work on which he had first cut his teeth as a naturalist, when working with Robert Grant. The delicate, skilled work would be a welcome change from writing about the geology of South America.

There was another reason for the enthusiasm with which Darwin set out on the task, which he thought might take him a year or two to complete. Although he had now been thinking about 'the species question' for several years, and the secret notebooks were bulging with ideas, he was uncomfortably aware that he had no reputation as a naturalist at all. The point had been very much hammered home when his new friend Hooker had written to Darwin in 1845, commenting disdainfully on a new book by a French writer who lacked the proper training:

> I am not inclined to take much for granted from anyone [who] treats the subject in his way and who does not know what it is to be a specific Naturalist himself.[2]

By a 'specific Naturalist', Hooker meant somebody who had made a firsthand study of taxonomy, and understood the nitty-gritty of how species were related to one another. Darwin very much took these comments to heart, even though Hooker had not meant them to be taken personally. The beloved barnacles would establish Darwin's credentials to theorise about the origin of species, and they would also – to an extent he could never have foreseen – provide him with the in-depth material he needed to flesh out his developing theory of evolution.

The snag was, in order to find out how these odd little creatures fitted in to the rest of the barnacle genus, Darwin needed to know what normal barnacles were like, and how they were related to one another. But the classification of barnacles was a complete mess. Nobody knew how normal barnacle species were related. As the project began to take shape, Darwin sought out barnacle specimens from museums and collectors near and far, eventually including specimens from all over the world. One of the best collections was in the possession of John Gray, the Keeper of Zoology at the British Museum. He had planned to make his own definitive study of the barnacle group, but had never found time; in 1847, as Darwin was beginning to realise the magnitude of the task facing him, Gray encouraged him to undertake the definitive study himself.

There was, indeed, nothing else for it. If Darwin was ever going to find the place of the species he had by now dubbed *Arthrobalanus* in the scheme of life, he was going to have to go the whole hog and sort out the barnacles properly. In all, it would take him eight years (although his ill-health lost him, by Darwin's own reckoning, two years' working time) and it would result in the publication of no fewer than four volumes on the barnacles, two describing living species and two describing fossil species.

The details of this epic study, which would alone have been enough to establish its author as a scientist of the first rank, need not bother us here – except for one extraordinary and fascinating insight which was to have a profound influence on Darwin's evolutionary thinking.

Barnacles are usually hermaphroditic, with each individual having both kinds of sex organ. But Darwin discovered first one, and then several, species in which there were two distinct sexes, united in a bizarre and previously unsuspected way. In the first of these species to be identified, which he called *Ibla cumingii*, Darwin found that all of the specimens he was examining (individuals about half an inch long) were females. What he had at first thought were parasites attached to the flesh of the barnacles turned out to be tiny males, whose sole function in life was to produce sperm with which to fertilise the eggs of the female. The males started life (like females) as independent, free-swimming larvae, but then attached themselves to a female, effectively as parasites, as Darwin had at first supposed.

In another species of *Ibla*, Darwin found that although the 'main' barnacle was indeed an hermaphrodite, it also carried tiny supplemental males. It looked as if this species was an intermediate form, in the process of discarding the hermaphroditic way of life and developing a sexual method of reproduction (in fact, it could be argued that the transformation happens the other way around, with males gradually shrinking away as hermaphroditism develops; but Darwin saw things as proceeding from hermaphroditism towards full sexuality). With his suspicions aroused, Darwin looked again at the 'parasites' on other species, and found several examples, ranging from fully hermaphroditic varieties through those with less developed male sexual organs and supplemental males to those with two distinct sexes, even though the males were, as he put it, 'mere bags of spermatozoa'. In some cases each female had a collection of 'little husbands', as he called them. In the end, he found that his original Chilean barnacle, *Arthrobalanus*, was itself a sexual species, with microscopic males attached to the shells of the tiny females – males in the form of a rudimentary head attached to an 'enormous coiled penis'.[3]

Through his classification and taxonomic work, Darwin not

only established the relationships between living and fossil forms of barnacles, but also their relationship with the rest of the crustacean community (before Darwin, nobody knew that barnacles were crustaceans; they were thought to be molluscs, relatives of snails rather than of crabs and lobsters). He showed that they were built on the same body plan as crabs, but with considerable modifications. When the juvenile form of a barnacle, which is a free-swimming larva, is ready to mature, it sticks itself to a rock (or some other surface) head first, and spends its adult life secure inside its shell, catching food with feelers that are highly modified versions of a crab's feet.

Both the relationship with other crustaceans and the extent of the modification of the body parts were key insights in the development of the theory of evolution. Darwin's intensive study of barnacles showed him that variety was not the exception in nature, but the rule – every part of every species could be found modified in some individuals, and there was no precise, unique body plan to which all members of the same species subscribed, only a broader outline. 'I have been struck,' he wrote to Hooker, 'by the variability of every part in some slight degree of every species when the same organ is *rigorously* compared in many individuals.'4

Darwin was clearly aware of the implications, although the theory of evolution was still only for private discussion with his trusted friends. He also wrote in delight to Hooker that:

I have lately got a bisexual cirripede [barnacle], the male being microscopically small & parasitic within the sack of the female . . . I tell you this to boast of my species theory, for the nearest & closely allied genus to it is, as usual, hermaphrodite, but I had observed some minute parasites adhering to it, & these parasites, I can now show, are supplemental males, the male organs in the hermaphrodite being unusually small, though perfect and containing zoosperms: so we have almost a polygamous animal, simple females alone being wanting. *I never should have made this out, had not my species theory convinced me, that an hermaphrodite species must pass into a bisexual species by insensibly small stages*; & here we have it, for the male organs in the hermaphrodite are beginning to fail, & independent males ready formed.[5]

But there was no mention of the 'species theory' when the epic work on barnacles was published. By the time the first volume appeared in print in 1851 (the final two volumes were not published until 1854) he was heartily sick of the project, and had begun to doubt whether it had been worth the effort that had gone into it. But the barnacle work was received with acclaim. Any lingering doubts were assuaged, and his flagging enthusiasm was revived when, at the end of 1853, the Royal Society awarded him its prestigious Royal Medal, with both his work on coral reefs and the first volume of his study of barnacles being cited. When the proposal for the award was moved at the Royal Society, Hooker told Darwin, there was 'such a shout of paeans for the Barnacles that you would have smiled to hear'.[6]

This was Darwin's coming of age as a biologist. He later described his career as a biologist as having three stages: 'the mere collector, in Cambridge &c.; the collector and observer, in the *Beagle* and for some years after; and the trained naturalist after, and only after, the Cirripede work'.[7] In his *Autobiography*, Darwin wrote that:

> the Cirripedes form a highly varying and difficult group of species to class; and my work was of considerable benefit to me, when I had to discuss in the *Origin of Species* the principles of a natural classification.

The work was of considerable benefit to others, too; Darwin's classification of barnacles is still the standard, definitive work. Thomas Huxley, then a young naturalist but soon to play a major part in promoting Darwin's theory of evolution, wrote that the barnacle study was:

> One of the most beautiful and complete anatomical and zoological monographs which has appeared in our time, and is the more remarkable as proceeding from a philosopher highly distinguished in quite different branches of science, and not an anatomist *ex professo*.[8]

After the barnacle epic, Darwin knew what species were all about from firsthand experience – and, crucially, his colleagues and contemporaries accepted him as an expert on species. That background would prove invaluable a few years later, when his species theory was at last aired in public. The theory of evolution is so important that we will devote a whole chapter to discussing it (Chapter 10); but even leaving that masterwork to one side for now, after establishing himself as a leading geologist in the 1840s and as a leading biologist in the 1850s, in the rest of his long life Darwin was to make several more contributions to the biological sciences, each of which would have been enough to mark the crowning achievement of a life in science for a lesser mortal.

There were also lesser contributions to science, although Darwin's enthusiasm for whatever his current topic of research might be never allowed him to treat any problem lightly, or to leave it without shaking everything he could from his studies. The first puzzle to catch his attention after the completion of the barnacle work was the question of how species colonised new lands – how they migrated there in the first place, and how they established themselves in competition with the native species when they got there. One possibility that Darwin (still a geologist!) toyed with was that the present-day continents of the southern hemisphere were all that remained of a former supercontinent – an idea with which he was, in many ways, a century or more ahead of his time. But the possibility he was able to test at home in Down was that similarities between the plants found in different parts of the world could be explained by seeds being transported by water.

Before Darwin, it had been generally assumed that seeds were killed by salt water, so that this method of spreading plant species would not work. But this was merely an assumption – nobody had bothered to check it. Darwin was different; he had no time for unfounded assumptions, and set out to find the truth. He took seeds of different varieties, and soaked them in salt water for different periods of time. Doing nothing by halves, he obtained seeds from far and wide and filled Down House with containers of soaked seeds, tended with loving care. And when he planted them, they sprouted.

Some seeds survived 42 days' immersion – long enough,

Darwin calculated, for them to float from Europe to the Azores, in Mid-Atlantic, on the prevailing currents. Others survived even longer immersions.

When confronted with the problem that although they would still germinate, seeds soaked in salt water sank, Darwin was unfazed. A spoonful of mud from the pond produced 29 plants, showing that seeds could be carried far and wide on the feet of birds; he collected birds' droppings to show that viable seeds could survive the passage through the digestive system of a bird, while being carried across the globe. And he even floated a dead pigeon for 30 days in salt water before recovering seeds from its crop and growing plants from them.

Where else would you publish the results of such experiments but in the *Gardeners' Chronicle*? Darwin did so in four articles in 1855, before presenting the results to the Linnean Society in 1856; they were mentioned, in due course, in the *Origin*.

His next interest played an even bigger role in determining the way Darwin's arguments would be presented in his masterwork. He needed to convince himself, and obtain evidence with which to persuade others, that new species really could be produced by the accumulation of small changes over many generations. He used pigeons to make his case, becoming so fascinated with the whole business of pigeon breeding and with the pigeon fanciers and their clubs that he went far beyond the strict requirements of his biological research in another of his many enthusiasms.

The variations that Darwin found among the pigeons bred by those fanciers were so great that any zoologist coming across them in the wild would have classified many of them as belonging to different species, some even to different genera. Yet those differences had been created in exactly the way Darwin had realised that natural evolution worked. The breeders obtained the results they wanted by breeding only from birds that showed the required characteristics, and by not allowing the others to reproduce. They selected the features they wanted. This was 'artificial selection', as opposed to the 'natural selection' that was soon to feature in the title of Darwin's masterwork. From pigeons, he moved on to studying rabbits and other domesticated animals, initiating a long project that would eventually lead to another major book, but not before he had been side-tracked by other interests.

The *Origin* itself, and the furore surrounding its publication (all of which we discuss later) took up most of Darwin's working time as the 1850s ended and the 1860s began. After the *Origin*, all of his books dealt with evolution and adaptation, and most of them dealt with plants, on which he could experiment himself (increasingly with the aid of his son Frank). For most of the latter part of his life, Darwin was a botanist as much as anything else – the world's first modern evolutionary botanist.

The first of these botanical fascinations was with orchids. It started on holiday in Devon in the summer of 1861, when Darwin spent hours absorbed in watching bees visiting wild orchids. This revived an old interest in the way insects are used by plants to ensure cross-pollination – the offspring of a cross between two different plants was, Darwin was convinced, stronger than the offspring of a self-pollinated plant (plants, like barnacles, are often hermaphroditic). But how did the plants ensure that the pollen they provided to the bees was used to fertilise the next plant the bee visited, and not the plant donating the pollen?

Darwin showed not only that the beautiful flowers of the orchid were actually there to guide the insects that pollinated them to exactly the right place to pick up the pollen while sucking their nectar, but also that the flowers were structured to ensure cross-pollination. Pollen was stuck to a visiting bee in such a way that it could not come into contact with the female parts of the same flower as the bee left, but must inevitably be brushed against the appropriate parts of the next plant that it visited. And not just any old insect could wander into an orchid's flower and do the trick; Darwin also showed how the elaborate structure of the flowers ensured that only specific species of insect could pollinate specific varieties of orchid.

In the same way that he had teased out the relationships between the barnacles and other crustaceans, Darwin identified the way in which the same basic plant structures had been modified by evolution to fulfil these specialised tasks. Had the orchid book appeared in 1852 rather than in 1862, it would probably have been seized on by religious orthodoxy as proof that these incredible structures had been designed by an intelligent being; coming three years after the *Origin*, however, the book, glorying in the title *On the Various Contrivances by which British and Foreign Orchids are Fertilised by Insects, and on the Good Effects of Intercrossing*, was another weighty piece of evidence in favour of evolution. It was also another major success in scientific circles, and sold a few thousand copies to a wider audience of orchid fanciers and a general public now curious about anything to which the name Charles Darwin was attached. *Orchids* remains one of Darwin's most readable books, as the following conclusions taken from it show:

> It is an astonishing fact that self-fertilisation should not have been an habitual occurrence. It apparently demonstrates to us that there must be something injurious in the process. Nature thus tells us, in the most emphatic manner, that she abhors perpetual self-fertilisation ... may we not further infer as probable, in accordance with the belief of the vast majority of the breeders of our domestic productions, that marriage between near relations is likewise in some way injurious – that some unknown great good is derived from the union of individuals which have been kept distinct for many generations?[9]

The reference to breeders of domestic varieties of animal touches on Darwin's main scientific interest throughout most of the 1860s. But his work on the variation of animals and plants under domestication proceeded in fits and starts, handicapped by his recurring illnesses. The botanical work, on the other hand, he regarded as something of a holiday, a hobby to which he could return as the mood took him.

The next enthusiasm was for climbing plants. Studies of hops, clematis and others showed Darwin how leaf stems had been modified by evolution to become hooks with which to cling on to supports, while leaf stalks had become elongated into tendrils. In 1867 the study appeared in the *Journal* of the Linnean Society, and in 1875 it was republished as a 118-page monograph.

Soon after his work on climbing plants began, in 1864, Darwin had been awarded the highest honour of the Royal Society, the

Copley Medal. And not long after the paper on climbing plants appeared in print, *The Variation of Animals and Plants under Domestication* was at last published, in 1868. It had taken seven years of intermittent work, and proved to be Darwin's longest book – it appeared in two volumes, each larger than the *Origin*. Once again, Darwin was convinced that it was not worth the effort he had put in to it, but the reaction from the scientific community proved that his fears were unfounded.

The history of *Variation* actually went back even further than seven years. In the middle of the 1850s, after he had completed his work on barnacles, Darwin had intended to spend several years organising his material on evolution and turning it into a very large book, which he planned to call *Natural Selection*, for later (perhaps even posthumous) publication. These plans were thrown into the melting-pot by the arrival of the famous letter from Alfred Russel Wallace, which revealed that he had independently drawn up the same theory of evolution that had occupied Darwin for so many years. The full story of the parallels between Darwin and Wallace are discussed later, in Chapter 10; what matters here is that the book we know as *The Origin of Species* was prepared hurriedly (by Darwin's standards) following the arrival of Wallace's letter, and was much shorter than the planned work on *Natural Selection*, intended only as a kind of summary or epitome of the great work.

In many ways this turned out to be all to the good, because it meant that Darwin presented his ideas in a much more accessible and readable form, suitable for a much wider audience than the epic scientific tome he was contemplating. But it did mean that he was left with a great deal of material that he had been gathering for the large book. In the early 1860s, his private notebooks make clear, Darwin was planning to turn this material into no fewer than three further books in support of his evolutionary ideas One of these, based on the material gathered for just the first two chapters of *Natural Selection*, became *Variation*; the other two books (*Variability of Organic Beings in a State of Nature* and *The Principle of Natural Selection*) were never written.

Unlike almost all of Darwin's other work, *Variation* was based largely on his interpretation of work carried out by other people, although his own work on pigeons features strongly, and some of his lesser studies get a mention. The pigeon studies are the archetype of what *Variation* is all about – an analysis of the way inheritance works in domesticated species, the 'artificial selection' that provides insight into the way natural selection works. Darwin's own ideas about heredity (again, discussed more fully in Chapter 10) were based on the idea that offspring acquired a blend of characteristics from both their parents. This was completely wrong, and in one of the great coincidences of science an obscure Moravian monk named Gregor Mendel was carrying out experiments which pointed towards the true nature of heredity at exactly the same time, in the 1860s, that Darwin was working on the *Variation*.

But although his understanding of the mechanism of heredity was wrong, *Variation* still provided what some modern commentators have referred to as 'a goldmine of particular information on domestic plants and animals'.[10] As well as the famous pigeons, it deals at length with artificial selection applied to such creatures as the turkey and other fowls, goldfish, and domesticated bees, as well as with cereals and vegetables. Discussing artificial selection (or 'selection by man', as he put it) Darwin wrote:

> When animals or plants are born with some conspicuous and firmly inherited character, selection is reduced to the preservation of such individuals, and to the subsequent prevention of crosses . . . but in the great majority of cases a new character, or some superiority in an old character, is at first faintly pronounced, and is not strongly inherited; and then the full difficulty of selection is experienced. Indomitable patience, the finest powers of discrimination, and sound judgement must be exercised during many years.[11]

That final sentence could be read (as Darwin clearly intended) as a summary of the way natural selection works, especially when taken in conjunction with his comment later in the same volume that 'the possibility of selection coming into action rests on variability'.

Variation was yet another success both with the experts and the public, with a first printing of 1,500 copies sold in a week, and a further 1,250 copies printed immediately to meet the demand. The book was published in January 1868, a few weeks before Darwin's 59th birthday, and it might be imagined that by now the sage was ready to take things easy and contemplate a well-earned rest. But not a bit of it. As ever, Darwin had become fed up with *Variation* by the time it was published, but as ever he was buoyed up by its success and off on his next project immediately.

The end product of this next phase of work was something of an oddity – *The Descent of Man, and Selection in Relation to Sex*, a book which is, as its title implies, really two books in one. *The Descent of Man* itself (the short title by which the combined work is often known, although only a third of the whole) was very much a sequel to the *Origin*, putting humankind in its evolutionary place. *Selection in Relation to Sex*, on the other hand, was really a supplement to the *Variation*, although Darwin did cleverly weave the two topics together in chapters that describe sexual selection at work in people.

The reason why the two topics were linked by him in one book is that Darwin believed (mistakenly, in the light of modern research) that sexual selection – the preference of partners for different notions of 'handsomness' or 'beauty' – was the main reason why the human species had split into different varieties, or races. We shall treat the book as the two different studies that it really is, saving the *Descent* for Chapter 12, and mentioning only the content of *Selection in Relation to Sex* here.*

The point about sexual selection is that beauty is in the eye of the beholder. It is indeed an important mechanism in evolution, although Darwin placed too much emphasis on it (largely because he lacked the knowledge of the way heredity works – essential for a full understanding of evolution at work). The classic example, used by Darwin himself, is the gorgeous tail of the peacock, which seems to serve no useful purpose, and may actually be a handicap to the male bird, making it harder

* Random House of New York has published a superb single-volume edition incorporating both *The Origin of Species* and *The Descent of Man, and Selection in Relation to Sex* in their Modern Library series. This is the best single volume summary of evolution by Darwin himself now available.

for him to escape from predators. How could such a seemingly useless adornment have evolved?

The sexual selection argument says that all that is needed to produce the peacock's tail is for pea*hens* to like it. It doesn't matter *why* they like it, although several suggestions have been put forward. Perhaps the pattern on the tail, like a ring of staring eyes, somehow hypnotises the female birds into submission. Perhaps, in some rather mysterious way, the females recognise that a male who can survive with the burden of a large tail must be a pretty strong character, ideal material for fatherhood. But whatever the reason, imagine that long ago peacocks had much more ordinary tails, but that peahens showed a preference, when choosing a mate, for males with slightly longer and more brightly coloured tails than average. The propensity for longer, brighter tails would be passed on to the next generation, and over many generations the tails of the peacocks would become larger and gaudier, until they reached their present day splendour. In Darwin's own words:

A girl sees a handsome man, and without observing whether his nose or whiskers are the tenth of an inch longer or shorter than in some other man, admires his appearance and says she will marry him. So, I suppose, with the pea-hen.[12]

In this way, Darwin explained the colourful beauty and exotic plumage of humming-birds and other brightly coloured species – another blow against the argument from design, which had held that such seemingly useless beauty in nature hinted at the work of a Designer who was not only intelligent but had a highly developed aesthetic sense. The bird of paradise did not get its beauty from a Designer, said Darwin, but from the mundane need to attract a mate.

It cannot be supposed that male Birds of Paradise or Peacocks, for instance, should take so much pains in erecting, spreading and vibrating their beautiful plumes before the females for no purpose.[13]

The *Descent* was published in 1871, just after Darwin's 62nd birthday, in two fat volumes. Once again, the first printing (this time of 2,500 copies) proved insufficient to meet the demand, and further printings brought the total to 7,500 copies before the end of the year. Darwin's next book followed hot on its heels, for the simple reason that it dealt with a subject that he had planned to include as a chapter in the *Descent*, but which grew too big to be dealt with in such a short space. This book, *The Expression of the Emotions in Man and Animals*, appeared in 1872, and was a sensation in Victorian Britain.

To a modern reader, it is hard to see why Darwin did not include this material with the *Descent* proper as one book, while separating out the material on sexual selection as a separate book in its own right. But it is easy to see why *The Expression of the Emotions* proved to be the most immediately successful of all Darwin's books, with more than five thousand copies in the bookshops (according to his *Autobiography*) on the day of publication.

Darwin's interest in how emotions are expressed by the muscles of the face went back to his days as a medical student in Edinburgh, and he also drew upon notes he had made of observations of his own children (especially William, the firstborn) in infancy.* He showed in his book that in many ways people react to stimuli in the same way that other animals, especially mammals, do, and he described what he regarded as the only uniquely human expressive emotional response, blushing, in terms that were decidedly racy by the standards of the day. On top of that, the book was full of interesting illustrations. No wonder it was a hit; many people probably found the chapter on blushing worth the price of the book on its own:

> Dr Browne, together with his assistants, visited her whilst she was in bed. The moment that he approached, she blushed over her cheeks and temples; and the blush spread

---

* These notes also formed the basis of a paper published in 1877 in the second issue of the first-ever psychology journal, *Mind*, putting Darwin at the forefront of yet another scientific discipline.

quickly to her ears. She was much agitated and tremulous. He unfastened the collar of her chemise in order to examine the state of her lungs; and then a brilliant blush rushed over her chest, in an arched line over the upper third of each breast, and extended downwards between the breasts, nearly to the ensiform cartilage of the sternum. This case is interesting, as the blush did not thus extend downwards until it became intense by her attention being drawn to this part of her person. As the examination proceeded she became composed, and the blush disappeared; but on several subsequent occasions the same phenomena were observed.[14]

In case you are in any doubt, let us assure you that this is indeed a description of a medical examination, reported to Darwin by Dr Crichton Browne, and not an extract from a Victorian bodice-ripper.

Now in his sixties, Darwin showed no sign of slowing his scientific output. With all his major works on evolution behind him, and with a marked improvement in his health (quite possibly because he had got evolution off his chest at last), he was able to indulge himself fully in pursuing his botanical hobbies. Like his study of orchids, the next book was a result of observations initially made on holiday. In fact, the first of these observations predated the orchid work, and were made in Sussex in the summer of 1860, where Darwin first became interested in the habits of insectivorous plants. He carried out experiments off and on throughout the 1860s, tempting such plants with various tit-bits and studying their responses, and came back to them after he had finished *The Expression of the Emotions*. The resulting book, *Insectivorous Plants*, was published in 1875, and is another good read, even today, for anyone interested in natural history. Its appearance had been delayed by Darwin taking time out to prepare revised editions of both the *Descent* and *Coral Reefs*, and in the background to all this he had been busy experimenting once again with the way plants reproduce.

In the mid-1870s, approaching his own 70th year, as well as preparing new editions of *Variation* and *Orchids*, Darwin turned these latest botanical studies into yet another new book,

The Effects of Cross and Self-Fertilisation in the Vegetable Kingdom, which was published in 1876. It was his biggest book on plants, based on a vast number of experiments and many painstaking observations. The whole thing stands as an archetypal example of how to carry out such scientific experiments, and provides overwhelming proof that cross-pollination produces stronger and more viable offspring than self-pollination. Darwin also came tantalisingly close to uncovering the important features of the hereditary mechanism that Mendel was investigating at about the same time, using the same kind of experiments but (probably crucially, as we shall see in Chapter 14) different kinds of plant, in which the hereditary patterns happen to show up more clearly.

The whole thing would have been a fine swan-song for a scientist approaching his 68th birthday, but Darwin was not finished yet. Another book about flowers appeared in 1877 (*The Different Forms of Flowers on Plants of the Same Species*), and a further investigation of the way plants move led to *The Power of Movement in Plants* in 1880. Surely, at the age of 71, Darwin had come to the end of his scientifically productive life? By no means. As the 1880s began, Darwin was hard at work on another enthusiasm, a study of earthworms and the way they affected the environment. He would publish ten scientific papers and a new book, *The Formation of Vegetable Mould, through the Action of Worms, with Observations on their Habits*, in 1881, the last full year of his life, and five papers in 1882, the year he died at the age of 73.

Earthworms is an excellent epitaph, containing all the Darwinian ingredients – careful observation, painstaking evaluation of evidence, and accurate, insightful conclusions. It built on an interest that had been stimulated back in 1837, when Josiah Wedgwood, Emma's father, had drawn his attention to the amount of soil brought to the surface in worm casts; over the intervening years, Darwin had published several short papers on the activity of worms. Now, he showed how important earthworms are in turning over the surface layer of the ground and creating topsoil, marvelled at their ability to bury stones and other objects by gradually undermining them, and pointed out that 'in many parts of England a weight of more than ten tons [10,516 kilograms] of dry earth annually passes through their bodies and is brought to the surface on each acre of land'.[15] 'Worms,' said Darwin, 'have played a more important part in the history of the world than most persons would at first suppose.'

The book sold 3,500 copies in two months, and in the first few years after its publication it sold better than any of his other books had in the first few years of their own life, including the *Origin*. Perhaps, like the posthumous hit records that often follow the deaths of musicians today, this success was partly a reflection of the sense of loss that people felt when Darwin died. But it is still a quite noteworthy, and largely unsung, achievement for a man whose name is usually linked today only with a book that had been published more than twenty years earlier.

Just as there was far more to the science of Albert Einstein than the theory of relativity, so there was far more to the science of Charles Darwin than the theory of evolution. Even without evolution, Darwin would have been one of the great nineteenth-century biologists; even without biology, he would have gone down in history as a great geologist. It is a measure of the importance of the theory of evolution that those other achievements seem modest in comparison to it; and it is time now for us to look first at the background to the publication of that theory, and then at the theory itself.

Notes

1, 2, 4, 6, 8 Quoted by John Bowlby, *Charles Darwin*, Hutchinson, London, and Norton, New York, 1990.

3, 12 Quoted by Adrian Desmond and James Moore, *Darwin*, Michael Joseph, London, 1991.

5, 7 Quoted by Ronald Clark, *The Survival of Charles Darwin*, Weidenfeld & Nicolson, London, 1985 [our italics].

9 Charles Darwin, *Orchids*, John Murray, London, 1862.

10 Duncan Porter and Peter Graham, *The Portable Darwin*, Penguin, London, 1993.

11 Charles Darwin, *The Variation of Animals and Plants under Domestication*, John Murray, London, 1868.

13 Charles Darwin, *The Descent of Man, and Selection in Relation to Sex*, John Murray, London, 1868.

14 Charles Darwin, *The Expression of the Emotions in Man and Animals*, John Murray, London, 1872.
15 Charles Darwin, *The Formation of Vegetable Mould, through the Action of Worms, with Observations on their Habits*, John Murray, London, 1881.

Chapter 9

Back to Evolution

The 1850s represent the era during which the British Empire began to be perceived as all-powerful, both in terms of its industrial might and its military power. By the beginning of the decade Britain was the wealthiest nation on earth and ruled the largest empire the world had ever known, turning the map pink. And, if the Great Exhibition of 1851 was anything to go by, the nation was not shy in displaying its evident success. The Great Exhibition simultaneously served two purposes: first it demonstrated to the world just how advanced and sophisticated a phenomenon British industry had become; at the same time it acted as an advertisement to the world, a clarion call to invest in Great Britain.

Charles Darwin was immensely impressed with the Great Exhibition. It opened on 1 May and was housed in a vast glass construction, the Crystal Palace in Hyde Park. The Palace itself was both a staggering engineering achievement and a great work of art which was tragically destroyed by fire in 1936. Darwin had read with interest reports about the Exhibition in *The Times* and friends and associates had waxed lyrical about it, but he and Emma were so preoccupied at Down that it was not until two months after the Exhibition had opened and had been visited by hundreds of thousands of people that they got to witness the event firsthand in July 1851.

The spring of 1851 had been a terrible time for the family. Still devastated by the loss of their beloved daughter Annie, Charles and Emma hardly had time to mourn before their ninth child, Horace, was born on 13 May, only three weeks after Annie's death. In the circumstances, and with Emma now in her early forties, it was a miracle that the pregnancy continued normally and both mother and child survived. Charles and Emma appear to have faced the situation stoically and although, for Charles at least, his daughter's death was the greatest tragedy of his life and he never fully recovered emotionally, the couple were helped by the fact that they had seven other children to support.

For their part the children reacted to the death of their sister in different ways. At eleven, seven and five respectively William, Henrietta and George were certainly old enough at least partly to understand what had happened, but Elizabeth, Francis and Leonard were far too young. If there is any validity in the theory that childhood trauma greatly influences future physical health, then it would appear that Henrietta, or Etty as she was known by the family, was the most seriously affected of the siblings, causing her parents great anxiety by suffering a series of illnesses which came and went over a period of years. As we will see, Darwin attributed this sickness to a genetic disorder in his children, but there may be some truth in the possbility that for Etty the loss of her closest sister at an impressionable age aggravated an already weak immune system.

Darwin not only saw the Great Exhibition as a marvellous demonstration of scientific advancement, but, coming from a family that had vicarious close associations with industry via the Wedgwoods and having grown up in an area which was fast becoming the country's industrial heartland, he was quick to realise the economic significance of British technological supremacy. Although the popular view of Darwin is that of a gentleman scientist locked away in his hermitage of Down House, absorbed by his experiments, the contents of his library and in his solitary walks between bouts of ill-health, this is only one aspect of his life and character. Darwin was certainly reclusive for some parts of his life, but although held back at the turn of the decade by his tragic loss, a few years later he began to blossom again and the 1850s may be seen as a decade during which he created his most acclaimed and famous work from the notes and mental images he had been gathering during the previous two decades. As well as this, Darwin was a very practical man who was keenly interested in making and keeping a comfortable life for himself and his family, a man acutely aware of the value of money, the need to maintain a social position and extremely anxious about the state of his economy, the education of his children and the condition of his investments.

As discussed in Chapter 7, in the mid-1840s Darwin had already begun to invest in land and was interested in earning as much as he could from his property. Although he was what today would be considered a liberal and believed in fairness and the just treatment of those less wealthy than him, he was nonetheless a very shrewd businessman. During the 1850s he invested heavily in industry, the railways and in land and spent a good deal of time away from his work supervising these investments. He was forever anxious about money. Robert Darwin had been a very knowledgeable and successful businessman and had lent large sums of money to hard-up members of the nobility, including a mortgage to the Earl of Powis, which at the time of Dr Darwin's death was worth £13,000. In all, when his father died, Charles inherited assets worth some £40,000 and the Wedgwood wealth contributed by Emma was in the region of £25,000, making the Darwin family a considerable financial force. Despite this wealth Charles was fretting constantly about the family's future security – an unusual attitude given that he had always known comfort and wealth and, unlike many who are born poor and acquire large amounts of money from their labours, he certainly could not have been fearful of returning to an impoverished past.

Like his father, Charles invested well; and despite his fears the family fortune grew considerably during the following years, mirroring the ascent of the country's economy. By investing in industry and booming technological developments, Charles was investing in the country and his family's fortunes climbed with those of the nation. This involvement in industry produced another spin-off: it gave Darwin a further intellectual paradigm for evolution.

By the early 1840s he had already developed the notion that nature was not the beautiful, peaceful rose garden, as suggested in the anthropocentric fantasies of the Reverend Paley and his ilk. Darwin had long since formulated the idea that a bloody struggle was continuously acted out in Nature, giving rise to the mechanism of natural selection as the driving force of evolution. Now, reflected in the pool of metal and steam, the might of the Great British Empire, and the turbines, blood and sweat that sustained it, Darwin could see another force – the need for the division of labour, the filling of niches in society, the principle that an industry can succeed only if a need is there or one can be created for it. In the same way, he realised, Nature lent strength to species which fitted their niches and allowed to develop the adaptation which most suited the environment. It was in this way that natural selection acted and species evolved.

At the beginning of the 1850s Darwin was still in no position to do very much with his evolutionary ideas, but things were changing. First, by 1854 his work on barnacles was reaching a conclusion and the last volume of his multi-volume epic was finished and delivered for publication. This came not a moment too soon. By as early as 1852 it was clear that Darwin was utterly sick of the creatures and wrote to his cousin William Darwin Fox: 'I hate a Barnacle as no man ever did before, not even a sailor in a slow moving ship.' Second, and more crucial, the intellectual *Zeitgeist* was altering for the better; soon at least some sections of the academic world would be ready to accept the validity of Darwin's ideas.

Spurred on perhaps by the advancement of British industry, a growing optimism in the future and the apparent success of the Victorian mechanistic approach, the political ferment of the 1820s and 1830s was replaced with a period of domestic stability, an atmosphere in which social reformers had a smaller, less motivated audience. At the same time a set of conditions was created in which new ideas were allowed to flourish; rebellion turned away from social change to intellectual change. Because Darwin was so anxious to see a shift in social awareness and a readiness for new intellectual developments, he was acutely observant of any sea change. As well as this, from his position high in the ranks of scientific London, he knew those involved with progressive thinking and later nurtured them for his own ends. In this we see another generally ignored aspect of Darwin's character. While working in isolation at home in Down House, he was busy manipulating the setting into which he could drop his bombshell when the time was right. Although not quite Machiavellian in scope, Darwin's scheming and plotting to manipulate the conditions into which he could plant his theory with success is far from the popular hermit-sage image of the man.

In the vanguard of new scientific attitudes in the 1850s were those involved in a periodical called the *Westminster Review*. A medically trained publisher, John Chapman (with whom the writer George Eliot had an abiding infatuation while acting as his assistant), ran the paper from the Strand and attracted a circle of intellectuals who would in the next few years become important members of the scientific establishment; in so doing he spearheaded a shift in the status and development of science in Britain. The circle included the philosopher John Stuart Mill, the physiologist William Carpenter, Robert Chambers (who wrote an incredibly successful but flawed popular science book called *Vestiges*, discussed in Chapter 10) and most importantly, a young former ship's surgeon, one Thomas Henry Huxley.

Until the *Westminster* reviewers began to help change the way science was done in Britain, scientific research had been a very exclusive world. Almost all scientists were either wealthy gentlemen such as Darwin, living off private incomes and indulging their interests and talents in research, or else they were Oxbridge graduates who had climbed the establishment ladder to take up positions as professors at universities and teaching hospitals. As a brilliant academic who had been born the son of a lower-middle-class schoolteacher, Huxley especially hated the Oxbridge system. He had graduated from Charing Cross Medical School on a scholarship and walked away with that year's gold medal for the final examinations. Coming from the other side of the tracks from the likes of Charles Darwin, he harboured a deep-rooted resentment for the class-ridden education system and the scientific establishment. For Huxley, merit could come only from intellectual ability and he was determined to smash the old system and to reform the blend of science and society. The logical future for the scientific establishment, he believed, was the creation of a well-paid scientific community. Scientists should be treated and paid like civil servants, they should be materially rewarded for their research without having to do it as a side-line to teaching. Research, he thought, should not be in the hands of crusty, tired old professors with sterile minds or the sons of wealthy land owners tinkering with science as a hobby. It was through this political zealousness that Huxley became a very useful tool for Darwin's campaign. At first, indeed until the publication of *The Origin of Species*, Huxley was deeply sceptical of natural selection as the mechanism for evolution and, in a similar way to Hooker, he constantly argued the matter with Darwin, offering him a much-needed counterpoint against which to sharpen his theory. When Huxley did become convinced of the truth of Darwin's ideas he, like Hooker, became an ardent supporter of evolution via natural selection. But, because of his crusade against the scientific establishment and in particular his loathing for certain sections of the scientific community who came out vocally against Darwin, Huxley took the matter much further than Hooker or any other of Darwin's supporters; he became an evolution evangelist, Darwin's orator and spokesman.

Back in 1852 as the *Westminster* reviewers were gathering, Darwin and Huxley had not yet met. Huxley had heard of Darwin's work and admired his books, but, totally unaware of Darwin's real stand and his then heretical ideas, he would most likely have viewed the naturalist as another target for his anti-establishment venom. For his part, Darwin was still straining over the microscope tweaking barnacles with his tweezers and writing up his findings, doing exactly what Huxley was most moaning about and at least two years away from returning to his work on evolution. He was also feeling horribly insecure.

He was ill again. The stress of Annie's loss had deep-rooted consequences which all his denial and intellectual suppression could not smother. He had a collection of worries to deal with. Although he was unaware of the problem, the preservatives in which the barnacles were kept are now thought to have adversely affected his immune system (see Chapter 5). Then there were his fears for the health of his children, especially the constantly sickly Etty, as well as concerns about the education of his sons and whether or not they would eventually take up respectable professions. He was even toying with the idea of emigrating to

Australia and wrote to his old assistant aboard the *Beagle*, Syms Covington:

> when I think of the future I very often ardently wish I was settled in one of our Colonies ... Tell me how far you think a gentleman with capital would get on in New South Wales.[1]

In voicing these notions, it is quite likely that Darwin was trying to run away from his problems – a characteristic he often displayed when the pressure grew too great. In a similar fashion he could never bring himself to visit Annie's grave in Malvern. In reply to a letter from Fox telling him that he had visited the grave, Darwin replied:

> Thankyou for telling me about our poor child's grave. The thought of that time is yet most painful to me ... About a month ago, I felt overdone in my work, and had almost made up my mind to go for a fortnight to Malvern; but I got to feel that old thoughts would revive so vividly that it would not have answered; but I have often wished to see the grave, and I thank you for telling me about it.[2]

He was tired of almost constant illness and sick to the back teeth of his work on barnacles, and then, late in 1853, his anxieties were multiplied by growing fears that war would break out in the Crimea. At first the war was seen by many as rather good fun, but when the wounded returned to England in their thousands the jingoistic fervour in which the Darwins, along with the majority of the nation, had become ensnared rang rather hollow. Although the Darwins were not directly affected by the war, cushioned as they were by wealth and privilege, the side-effects were plain to see. Young men were off to the front, the workforce was diminished and there was always the fear that the war could cause economic problems. In Darwin's overactive imagination, the war could curtail the blossoming of the young empire and bankrupt the country, taking his investments with it.

Of course, he was overreacting and his fears of financial ruin were those of a man weighed down with other, less nebulous fears, but nonetheless largely unable to face them head on. The Darwin boys played soldiers in the grounds of Down House and imprisoned the girls in one of the outbuildings, and some of the men of the village rode off never to return, but that was the extent of the family's involvement in the conflict. Yet, intellectually it served another purpose. For Darwin, the struggle in the Crimea was yet another model upon which to base his ideas of natural selection; it mirrored the war raging constantly in Nature, the battle for survival played out in every field, every lake and every pile of dust in every corner of every room.

At last, during the spring of 1854, after years in exile imposed by ill-health and suppressed grief, his work on the dreaded barnacle and the terror of society's reaction to his secret ideas, Darwin emerged from his cocoon, feeling well, enriched by a belief that the mood of the times was changing in his favour and he began to think again about evolution. A new, energised Darwin was ready to take on the world.

Even so, it took Darwin almost eighteen months to begin writing what he considered to be a publishable version of his theory. In this he was greatly encouraged by his old college mentor Charles Lyell. Lyell really was instrumental in coaxing out of Darwin what became *The Origin of Species*, because, although Darwin was beginning to sense that the time would soon be appropriate for some form of declaration of his views, he had no idea how he was going to present the theory. It was Lyell who pushed Darwin into writing a book rather than delivering a paper on the subject. Darwin was always grateful for this encouragement, as he expressed, for example, in a letter to Lyell of November 1856:

I am working very steadily at my big book; I have found it quite impossible to publish any preliminary essay or sketch; but am doing my work as completely as present materials allow without waiting to perfect them. And so much acceleration I owe to you.

In some ways Lyell's enthusiasm was surprising because he was still quite unsure about evolution and the concept of natural selection. Coming from an older generation and a Christian background he found it difficult to accept Darwin's theory for the simple reason that it had, by definition, to include Man. This was of course the great sticking-point with the entire theory. The notion that animals evolved could just about be contemplated, while the notion that new species are not created spontaneously or at the whim of a Creator could be considered a possibility; after all, it was suggested, perhaps a Divine Being created the simple forms from which more complex species evolved. But then arose the thorny matter of Man. Man is an animal: what is there to distinguish him from the other animals?; what is it about Man that gives him a special place in creation?; how could Man have an immortal soul if he has evolved from simpler life-forms? Darwin's answer would have been: Man is no different to any other animal; perhaps there is no such thing as a soul and the Old Testament tales of a Creation are mere fiction. In other words, a special place for Man might be just so much hocum. Naturally, there were many who could not contemplate this conclusion and Lyell was one of them.

Despite Lyell's encouragement for the project, Darwin appears to have had little sympathy for his old teacher's misgivings and concerns. Shortly before *The Origin* was published and after years of discussing the matter in letters between them, Darwin wrote to Lyell:

> I'm sorry to say I have no 'consolatory view' on the dignity of man. I am content that man will probably advance, and care not much whether we are looked at as mere savages in a remotely distant future.

Not only is this an interesting letter for the fact that Darwin takes such a hard line with his old mentor's sensibilities, but it is also striking that throughout the paragraph Darwin uses a small 'm' for Man – perhaps a further example of the extent to which he had totally lost faith not only in God, but in Mankind being in any way special.

Eventually Lyell did manage to find an answer that satisfied him, a solution which allowed for the dignity of Mankind as well as the validity of hard scientific evidence. In the meantime he was a good enough scientist to understand the value of Darwin's great work and was also aware that it was important for Darwin to present his work in published form simply to prevent anyone else beating him to it. Strangely, this notion appears never to have crossed Darwin's mind and when he was made aware of the possibility, he thought it a trivial reason to publish, declaring: 'I rather hate the idea of writing for priority, yet I certainly should be vexed if any one were to publish my doctrines before me.'[3]

At first Darwin planned to write a short book, but very soon the project took on a life of its own and ended up being a massive work. He began writing in May 1856 and in an initial burst of enthusiasm completed 40 pages by July before another domestic problem disturbed him. At the age of 48 Emma had again become pregnant.

Charles was extremely anxious for his wife. In the nineteenth century childbirth was a dangerous event and although Emma had already borne nine children, at her age anything could happen. His way of dealing with his worries was to plough on with his work, and with Emma the new invalid of the household it was his turn to supply comfort and attention between long stretches of writing. By the time Emma gave birth to their sixth son in December, Darwin was onto the third huge chapter of his book, which had already grown to some 150 pages in length.

The new Darwin was named Charles Waring and it soon became apparent that he was retarded. Although the reason for the baby's arrested development was unknown at the time, it has been suggested in recent years that he suffered from Down's syndrome.[4] Again, this latest medical problem sent Darwin into a spin and further fuelled his suspicions that his family may have been genetically weak because of the close family relationship between himself and Emma. 'My fear is hereditary ill-health,' he told Fox. And in reference to his children, goes on to say: 'Even death is better for them.'

Darwin believed that in his family's case the evidence was clear. He was suffering from a mysterious, recurring illness, Annie had died in childhood displaying many of the same symptoms with which he was afflicted, Etty and George were frequently ill, Elizabeth, their fourth child now approaching her ninth birthday, was an unusually quiet girl who suffered from learning difficulties, and now the latest child had been born abnormal. He and Emma were cousins and he had, he believed, passed on his medical weaknesses to his children because, by marrying a relative, his genetic material had not been given the opportunity to compete with perhaps stronger genes in the human gene pool.* As he wrote about inbreeding in the animal kingdom in his vast treatise, the evidence for his ideas was all around him, in his own home, intruding into his own life.

Darwin may have been alerted to the notion that some other scientist could even at that moment be working on a similar theory but it appeared to do little to change the speed at which he worked or his attitude towards the kind of book he was planning to produce. By 1857 he had decided that the book would be a highbrow account of his theory aimed at his scientific colleagues rather than the general public. In those relatively innocent days, almost three years before publication, Darwin's view of how his theory could be presented and who would read it were naive in the extreme. Darwin saw his audience as the intelligentsia. This, he believed, would be where the battle over his heretical notions would be played out upon publication, these were the people he would have to convince. Further evidence for the fact that Darwin was not at all worried about being superseded comes from the fact that at around this time, in early 1857, he appears to have rather rapidly run out of steam. The initial impetus which had sustained him for a year was already beginning to evaporate.

For the past five years he had been returning sporadically to his water treatment. For a few weeks at a time he would rise early and take cold dips, then give it all up for several months. In retrospect it is clear that the treatment did work. Whether this was because in some unknown fashion it acted in a physiological way or the obvious benefits Darwin gained from it were derived

---

* Although, as discussed in Chapter 12, Darwin did not know of the mechanism by which characteristics were inherited and the terms 'gene' and 'genetics' had not been coined by the 1850s, as the basic tenet for his theory he was aware of the fact that characteristics were passed on from generation to generation.

from a purely psychosomatic mechanism is of course impossible to prove. He either failed to see a close connection between better health and commitment to the treatment or else he did notice it but just could not face the treatment for long periods because of the unpleasant associations it held. Darwin was sufficiently aware of the nature of his illness to capitalise on periods of good health, such as the spell between the spring of 1856 and that of 1857 during which he worked flat out on the book. Based on bitter experience, he could never tell if or when sickness would return. Emma was acutely aware of how her husband's illness fluctuated with his mental state and stress levels, once claiming: 'Charles' health was always affected by his mind.'[6]

Even periods of good health had their down side – during these times he of course overdid it. Trying to make up for lost time, he ended up exhausting himself. This is exactly what happened during the early months of 1857. After a spell working at full steam Darwin found himself spiralling back into erratic health. Again this new bout was coupled with domestic problems. Ironically, despite his anxieties over Emma's condition, he had blossomed during her pregnancy with Charles Waring, but as soon as the child was born defective and his other children were continuing to fall prey to a succession of illnesses, Darwin again began to feel his own good health slipping away. The writing of the book slowed dramatically and this served to further compound his problems. It was the sheer size and complexity of the project which now weighed heavily upon him. A year into the book and he could hardly see an end to it. Once more it appeared that he had begun a project which would take far longer than he had predicted and might have no intrinsic value.

Taking himself off for two weeks to another water treatment establishment near Farnham in March 1857 appeared to do the trick for him, but little altered Etty's health; she had been taken to Hastings by Emma to recuperate at the same time. Charles' improvement was another temporary one. The summer of 1857 was unusually hot and the baby's christening in May and the houseful of visitors it entailed caused him further stress and he took to bed for several days. Work on the book ground to a complete halt. Then, when the worst was over, he pressed on as best he could, back to the old system of working for two or three hours a day when he was not feeling well, even this regularly interrupted by weeks of rest in Farnham; working at full speed when he felt better and exhausting himself to the point where he was forced back to bed once more. All the while, as the book grew in slow, painful stages, Darwin's other, equally important task was to lay the groundwork ready for publication, clandestinely nurturing key figures within the scientific commmunity who would be instrumental in the advancement of his theory.

Darwin was ideally placed for such a programme of intellectual coercion. He was a highly respected figure within the scientific community. With his acclaimed work on barnacles he had acquired all-important kudos as a biologist and he had friends in high places. Thanks to his voyage on the *Beagle*, he was regarded as a man of wide experience and he had already published several books on a variety of subjects mostly associated with his voyage. Darwin was fully aware of who the crucial people would be. He knew that he was not a great speaker and was not healthy enough to proselytise his theories in person; therefore, long before anything was ready for publication, he began to influence the thinking of Huxley, Hooker and a few other selected naturalists.

Darwin was worried about Huxley. The evangelical reformer was if anything going too far, too fast. After all it was Huxley who in 1859, and after receiving Darwin's calming influence claimed: 'If I have a wish to live thirty years, it is that I may see the foot of Science on the necks of her enemies.'[7]

He was not keen on Darwin's notions of evolution and natural selection, a bias which Darwin gradually corrected, but in some ways more importantly, Huxley was becoming too antagonistic within the scientific community. By 1857 he had been Professor of Natural History at the School of Mines and Museum of Practical Geology in central London for three years and had imbedded himself deep within the scientific establishment, a position from which he was vigorously attacking his hated targets. Top of his hit list was Richard Owen, whom he saw as the worst example of a pseudo-scientist, wrapping up his biological ideas in a neat anthropocentric mysticism. Huxley never missed an opportunity to attack Owen both in his lectures and in his writings for scientific journals and newspapers. Darwin saw this aggression as being too partisan and as his relationship with Huxley grew more intimate he took great pains to nurture in the younger man a broader understanding of the issues, encouraging him to be less focussed in his attacks and to widen his perceptions. At the same time Darwin was influencing Huxley into an acceptance of the basic tenets of natural selection. In this he can be seen as having been highly successful because, when *The Origin* was eventually published, Huxley was converted to the theory immediately and wholeheartedly. This construction of a supportive infrastructure for the theory was crucial when evolution via natural selection went public. Single-handedly and in his frail state of health, Darwin would have been quite unable to have defended adequately his theory.

By June 1858 Darwin's enormous book was nearing completion and the training of his front-line troops was moving along nicely. The weather was clement and for some weeks he had been feeling unusually well, enjoying his country walks and feeling pleased with the gargantuan effort now approaching a conclusion. Then, on the morning of 18 June, the postman arrived with a letter from a colleague of Darwin's living in the Far East. Moments later Darwin's whole world seemed to fall apart.

The letter was from a young naturalist, Alfred Russel Wallace, who at the time was travelling in the Far East collecting specimens and selling them to wealthy enthusiasts in Europe. The son of an unsuccessful solicitor, Wallace was a self-educated biologist of remarkable intelligence and insight. Having come from a poor background and surviving on his wits and determination, by the time he made contact with Darwin he was still only in his early thirties but had travelled the world working as a specimen-hunter, surveyor and teacher, had presented papers to a clutch of learned British scientific societies and had published an account of his travels. Through a combination of firsthand experience and reading both popular and serious scientific texts Wallace had, by 1858, come to a very similar conclusion to Darwin concerning the evolution of species. By a remarkable twist of fate, almost Darwin's social opposite, a man who had travelled a similar physical path to Darwin but in very different circumstances, had reached an almost identical conclusion. It was the announcement of Wallace's theory contained in the letter which so shook Darwin that, in the short term at least, his plans and careful schemes appeared to have been completely wasted.

The fact that someone else was developing a similar theory to Darwin's at the same time was not in the least surprising and besides, Darwin had been warned. What is surprising, at least with hindsight and from a dispassionate viewpoint, is the fact that, given the circumstances, Darwin had not realised that Wallace of all people was onto something.

Legend has it that Darwin woke up one day to learn of the existence of Alfred Wallace and his identical theory. This is wrong on several accounts. The most striking falsehood in this version is that, far from learning of Wallace's existence on that June morning, Darwin had known of him for at least two years and had been corresponding with him for over a year, he had even purchased a number of specimens from him and had had them transported to England from the Far East. Furthermore, Wallace was using Darwin as a sounding-board for many of his ideas about evolution. Wallace had read Darwin's books and held him in high esteem. Compared with Wallace in the 1850s, Darwin was a famous man – near the apex of the scientific hierarchy, a respected member of the scientific community, admired fellow of the Royal Society, acclaimed author and respected biologist. In other words a wonderful contact.

Darwin appears to have been deaf to the warning signs. As early as 1856 Lyell had tried to alert him when he mentioned a paper written by Wallace that had appeared in the *Annals and Magazine of Natural History*, a paper in which the author discussed the matter of the evolution of species but had carefully disguised his real meaning with a thin veil of Creationism. Many readers had been fooled, including Darwin, but Lyell had not been. Reading between the lines, Lyell had interpreted Wallace's argument as being close to what he knew of Darwin's line of thought. Darwin merely saw Wallace as using some of his own old ideas and blending them with Creationism. Because of this blind spot he continued to exchange ideas in his correspondence with Wallace almost as a patriarch, all the while unwittingly encouraging Wallace to come out in the open about his true beliefs. In an almost identical way to Darwin's approach with certain key characters in his acquaintance, Wallace was testing the water with Darwin to see if he could trust him with his own dark, heretical secret. When he felt he could expose the entire concept, he did, and with devastating effect.

As we will see in the next chapter, Wallace's ideas were not identical to Darwin's but they were so close that Darwin's immediate conclusion was that he could do nothing but concede originality to Wallace without a moment's hesitation. In a panic, Darwin was ready to hand over priority to Wallace and to step aside from any contest or argument. Overawed, he wrote to Hooker and Lyell, describing the situation and telling them that he could not deal with the matter and turning all responsibility over to them. To Lyell he wrote:

> I never saw a more striking coincidence; if Wallace had my MS sketch written out in 1842, he could not have made a better short abstract! Even his terms now stand as heads of my chapters. Please return me the MS, which he does not say he wishes me to publish, but I shall, of course, at once write and offer to send it to any journal.

There were two reasons for Darwin's mental state. The first was that Wallace's letter came as a shattering blow. It is difficult to appreciate just how devastating was the news. Darwin had been working on his theory for twenty years, it was his most cherished work and meant more to him than all the barnacles in Christendom, yet he was not the sort of character to argue over priority for a discovery. It had been lax, even remiss of him to have ignored the signals coming from the other side of the world, but it was through absent-mindedness, not arrogance, that he failed to perceive the danger. He was so wrapped up in his work that he could not spare the intellectual energy to ponder whether or not he would be first to publish. It was not until the threat of being usurped lay there in black and white that the full force of what it meant struck him. Darwin himself claimed not to have been concerned with whose name was put to the discovery of evolution by natural selection, but was nonetheless shattered by the fact that it then seemed that Wallace had swept away 20 years' work and that his own efforts had been a waste of that time. Clearest evidence to support this is that, upon hearing the news, Darwin's first reaction was to write to Wallace to congratulate him and to offer to help place his paper with a respected journal. It was only through the interference of Lyell and Hooker that he was stopped from doing this and eventually a happy outcome was achieved.

Another reason Darwin felt that he could not deal with the problem personally at that moment was a purely domestic one. On 28 June, a little over a week after the arrival of Wallace's letter, Charles Waring Darwin died from scarlet fever. Naturally Darwin was at the centre of an emotional whirlwind. On the one hand it appeared that his life's work had been pre-empted and on the other, his youngest child had died. Scarlet fever eventually claimed the lives of six children in the village and for the Darwins the loss of Charles Waring produced mixed feelings. Although they were naturally deeply saddened, clear as it had now become that the child was severely retarded, Charles and Emma could also view the loss as a blessing in disguise. Even so, the culmination of events was too much for Darwin to deal with at that moment. As the drama in Down House unfolded, his friends were dealing with the Wallace matter.

The solution they came up with was a joint announcement. After some successful stringpulling, Hooker and Lyell managed to book the recently revitalised Linnean Society, now operating from its new premises in Piccadilly, for a reading on 1 July, the last meeting before the summer recess. The reading of Darwin's theory was based on an outline of his 1844 essay and part of a letter from Darwin to the American botanist Asa Gray, and the paper Wallace had sent to Darwin was read immediately afterwards.

Even then, Darwin, still overly concerned that he was treading on Wallace's toes, was not entirely convinced that Lyell's and Hooker's solution was appropriate. Later in his autobiographical writings, he said of the incident:

> The circumstances under which I consented to the request of Lyell and Hooker to allow of an abstract from my MS., together with a letter to Asa Gray, dated September 5, 1857, to be published at the same time with Wallace's Essay, are given in the *Journal of the Proceedings of the Linnean Society*, 1858, p.45. I was at first very unwilling to consent, as I thought Mr Wallace might consider my doing so unjustifiable, for I did not know how generous and noble was his disposition. The extract from my MS. and the letter to Asa Gray had neither been intended for publication, and were badly written. Mr Wallace's essay, on the other hand, was admirably expressed and quite clear.

Much to Darwin's surprise, this first public declaration of his and Wallace's theories went almost unnoticed. Perhaps it was because the society had reached the end of a long working period and its members were ready for a break, but the reception the papers generated was muted to say the least. In retrospect, the now famous comment made by Thomas Bell, the President of the Linnean Society, lamenting the fact that that year's sessions of the society had 'not been marked by any of those striking discoveries which at once revolutionise, so to speak, [our] department of science'[8] now seems rather unfortunate, but Bell was not the only one who failed to realise what he had witnessed that evening. Perhaps Hooker's own recollection of the evening gives a better picture:

> There was no semblance of discussion. The interest excited was intense, but the subject too novel and too ominous for the Old School to enter the lists before armouring. It was talked over after the meeting, 'with bated breath'. Lyell's approval, and perhaps in a small way mine, as the lieutenant in the affair, rather overawed those fellows who would otherwise have flown out against the doctrine, and this because we had the vantage ground of being familiar with the authors and their themes.[9]

With the emergence of Wallace onto the scene and the gentlemanly resolution to what could have presented Darwin with an enormous problem, it was clear that Darwin had to finish his book and to publish soon. Wallace always acknowledged the fact that Darwin was the man to write the book describing the theory of evolution and dispensed gamely with any initial plans to start his own. Wallace also always gave full credit to Darwin as the primary discoverer of the theory and never contested the issue. For his part Darwin never failed to declare Wallace as joint discoverer.

Once again it was Lyell who pushed Darwin into publication. It was evident that Darwin's enormous text would be quite unsuitable for the new purpose of a popular account. With the reading of his and Wallace's papers at the Linnean Society there was now no real need for a high brow, academic treatise on the theory. What was really called for was a general account, a book which would appeal to the educated public, something both erudite and readable. Luckily, Darwin had already planned to write an abstract, a condensed version of the original book, and he began this almost immediately, during a family holiday on the Isle of Wight late in July 1858. By April of 1859, a couple of months after his 50th birthday, the manuscript was complete, a version pared down to a mere 155,000 words – what would eventually become the published version of *The Origin of Species*.

Throughout the autumn and winter of 1858 and through the spring of 1859, Darwin worked at full speed on the final version of his book. From time to time he would send chapters to Hooker and Lyell for correction and then, when the first complete draft was finally written, this too went off to his colleagues for final comments. To Lyell he wrote:

> I look at you as my Lord High Chancellor in Natural Science, and therefore I request of you, after you have finished, just to rerun over the heads in the recapitulation-part of the last chapter. I shall be deeply anxious to hear what you decide (if you are able to decide) on the balance of the pros and cons given in my volume.

Thanks to Hooker's children, the book very nearly never saw the light of day. One afternoon when Hooker was not paying them enough attention they dug out Darwin's manuscript from beneath a pile of books and used it to draw on. The book was saved only when Hooker realised belatedly what was happening and retrieved it before too much damage was done.

With the manuscript complete, it was again up to Lyell to push it through to the next stage. Acting as Darwin's adviser and literary agent he managed to secure a publication deal with John Murray, the publisher responsible for Lyell's own *Principles of Geology* and Hooker's *Himalayan Journals*. After months of proof-reading and further large-scale corrections, a period when Darwin could simply not leave his manuscript alone, it was finally ready for publication. During the final stages of preparation the book went through numerous title changes (see Chapter 10), Darwin's original – *An Abstract of an Essay on the Origin of Species and Varieties through Natural Selection* – eventually being shortened to *On the Origin of Species by Means of Natural Selection*.

In view of Darwin's earlier literary successes, it is perhaps surprising that the publishers were initially unconvinced that the book would be a commercial success, and it is quite likely that it was accepted only because of Darwin's fame within the scientific community and the quality of his supporters already known to the company. Even more surprising is the fact that, thanks to Lyell's reassurances to the publishers that the book was not blasphemous, Murray agreed to take it without first reading the manuscript, a decision which in commercial terms they were never to regret. Ironically it is quite possible that for all Lyell's assurances, they would otherwise not have accepted the book, as Mr Murray had himself rejected a book by the acclaimed novelist Harriet Martineau on the grounds of its 'infidel tendency'. The lack of enthusiasm for the project can be seen in the fact that John Murray anticipated a first print run of only 500 copies. But, as the publication date approached, the projected sales figures began to grow. By the time it was ready for the shops, 1,250 copies had been printed and the publishers were by then so keen on the project that they even agreed to absorb the bill for £72 to cover the enormous number of late corrections and alterations which Darwin insisted on making to the manuscript.

Darwin was not so enthusiastic. The moment of truth

was almost upon him. At the beginning of October, with the last alterations completed, the final proofs checked and double-checked, Darwin packed himself off to a hydropathic therapy home in Ilkley in Yorkshire. Here he sat and brooded over the outcome of the publication of his book. For over two decades he had held back from announcing his ideas to all but a few trusted colleagues and his wife Emma and such efforts had considerably aggravated his illness. Now there was no turning back. Of course the book might fail to make an impression, just as the reading of his and Wallace's papers had failed to generate a reaction at the Linnean Society meeting the previous year. Secretly Darwin half hoped that this would indeed happen. Needless to say it did not.

Notes

Unless otherwise specified, quotations in this chapter are from *The Autobiography of Charles Darwin and Selected Letters*, 3 vols, ed. Francis Darwin, John Murray, London, 1887.

1, 3 *The Correspondence of Charles Darwin*, 8 vols, ed. F. Burkhardt and S. Smith, Cambridge University Press, 1985–93, vol. 4, p. 362

  2 Letter to Fox, 3 October 1856, from the unpublished letters of C. Darwin to W. D. Fox in the library of Christ's College, Cambridge.

  4 John Bowlby, *Charles Darwin: A New Biography*, Hutchinson, London, and Norton, New York, 1990, p. 329.

3, 5, 9 *The Life and Letters of Charles Darwin*, 3 vols, ed. Francis Darwin, John Murray, London, 1885.

  6 Quoted by Adrian Desmond and James Moore, *Darwin*, Michael Joseph, London, 1991.

  7 Letter from T. Huxley to F. Dyster, 30 June 1859, from the T. Huxley Papers at the Imperial College of Science and Technology, London.

  8 *Journal of the Proceedings of the Linnean Society*, Zoology II (1858), p. 45.

  9 *Life and Letters of Sir Joseph Dalton Hooker*, 3 vols, ed. L. Huxley, Macmillan, London, 1918.

Chapter 10

The Masterwork

Darwin had begun thinking about 'the species problem' long before the end of his voyage on the *Beagle*. In his *Autobiography*, he described how during the voyage he had been deeply impressed:

> by the manner in which closely allied animals replace one another in proceeding southwards over the [South American] Continent; and ... by the South American character of most of the productions of the Galapagos archipelago, and more especially by the manner in which they differ slightly on each island of the group.

What Darwin had noticed, in modern terms, is that many of the species on the Galápagos (birds, for example) had evolved from a common ancestral stock and adapted to conditions on each island. In one place the birds might have long beaks for probing into crevices, while on another island the close relations of those birds, descended from the same ancestor, had evolved heavy beaks for crushing tough seeds. 'Such factors as these,' wrote Darwin in his *Autobiography*, 'could be explained on the supposition that species gradually become modified.'

He began to work seriously on the puzzle of how this could happen once he was settled in Great Marlborough Street in 1837. The first of the famous notebooks was started in July that year, and he began to pester anyone who might have relevant information with questionnaires. The biographer Ronald Clark describes one of these documents, preserved among Darwin's papers – an eight-page pamphlet consisting of 21 numbered paragraphs listing 44 specific questions, beginning with:

> If the cross offspring of any two races of birds or animals, be interbred, will the progeny keep as constant, as that of any established breed; or will it tend to return in appearance to either parent?[1]

Darwin already had little or no doubt that species did change. But the key question facing him as he began his first notebook on the species problem, and sent out his questionnaires, was *how* evolution could occur. The insight which led to the theory of evolution by natural selection and the notion of 'survival of the fittest' came to him after he read the famous *Essay on the Principle of Population*, by Thomas Malthus, in the autumn of 1838.

This was actually the sixth edition of that work, which had first been published anonymously in 1798, when Darwin's grandfather Erasmus was still alive. Malthus himself had been born in 1766, studied at Cambridge University and was ordained in 1788. Unlike many of his contemporary academics, Malthus actually carried out religious duties for a time, working as a curate at Albury in Surrey. He later became Professor of History and Political Economy (Britain's first) at Haileybury College, run by the East India Company; but it was while working as a curate in the 1790s that he produced the first version of his *Essay*. He only later became known as a leading economist, and was elected a Fellow of the Royal Society in 1819. He died in 1834, while Darwin was still only a ship's naturalist.

Malthus was intrigued by the way in which populations, including human populations, had the potential to increase at a geometric rate of growth. This means that the population doubles in a certain interval of time, doubles again in the next interval the same length, and so on.

The power of such a geometric progression is familiar from the story of the sage who did a favour for a potentate in ancient times, and asked as his reward a modest amount of grain –

one seed for the first square on a chessboard, two seeds for the second square, four for the third square, and so on. The potentate happily agreed to his wish – not realising that the 64th and last square of the board would be associated with 2^{64} grains of corn, roughly equivalent to the number 18 followed by 18 zeroes. If each grain of corn represented a second of time, this number would be equivalent to about 30 times the age of the Universe in seconds, according to the well established Big Bang theory. And this huge number has been achieved with only 64 doublings.

Imagine a world in which every wind-blown dandelion seed survived to grow into a new plant, or in which every egg in every mass of frogspawn developed to be an adult frog, and you have some idea of the power of geometric progression People reproduce more slowly than frogs, or dandelions; but Malthus pointed out that in the new lands of America at that time population was actually doubling rather more quickly than once every 25 years. At that rate, in 64 × 25 years (only sixteen centuries!), human population would reach the ridiculous 'chessboard number', 18 followed by 18 zeroes.

At one level, this capacity for reproduction is not surprising. All that is required is that each couple should have enough children, by the age of 25, to ensure that four of those children survive to have children in their turn. On the other hand, it is quite clear that human population has not doubled every 25 years since time immemorial, or even since the time when Empedocles was speculating about the origins of species. Even such a slow-breeding animal as the elephant could, starting from a single pair, produce a population of millions of elephants in less than a thousand years if each pair produced four offspring that survived to reproduce in their turn. But the world is not overrun with elephants, or frogs, or dandelions, or anything else. On average, each pair of elephants that was around a thousand years earlier had left one pair of descendants in Malthus' time. Leaving aside the influence of 'civilised' humans, natural populations stayed more or less constant. Why?

Malthus realised that all populations are held in check by restraining influences, including the actions of predators, sickness, and especially the limited amount of food available.

In everyday language, population expands *only* far enough to consume the resources available. In his own words:

> The natural tendency to increase is everywhere so great that it will generally be easy to account for the height at which the population is found in any country. The more difficult, as well as the more interesting, part of the inquiry is to trace the immediate causes which stop its further progress ... What becomes of this mighty power ... what are the kinds of restraint, and the forms of premature death, which keep the population down to the means of subsistence.[2]

The message that Malthus' *Essay* brought home to Darwin is that the *majority* of all individuals, in the natural state, do not survive long enough to reproduce. Darwin wondered why some individuals, the minority, should survive and reproduce, while others did not. And he saw that the survivors would be the ones best suited to the way of life of that species – best fitted to their ecological niches, in the way that a key fits into a lock, or a piece fits into its place in a jigsaw puzzle.

On the day that he read Malthus' *Essay*, Darwin wrote in his own 'Notebook on Transmutation of Species':

> On an average every species must have same number killed year with year by hawks, by cold, &c. – even one species of hawk decreasing in number must affect instantaneously all the rest. The final cause of all this wedging must be to sort out proper structure ... there is a force like a hundred thousand wedges trying to force every kind of adapted structure into the gaps in the economy of nature, or rather forming gaps by thrusting out weaker ones.[3]

The image of 'a hundred thousand wedges' is one that Darwin was fond of, and it would recur in his later notes and his published work on evolution. But here, in the autumn of 1838, all the essential ingredients of the masterwork are present – the pressure of population, the struggle for survival, and the survival of the fittest. Before Darwin, other people had realised that there is competition in nature; but they had seen any such 'struggle' as between different species, competing with one another – like different football teams competing for a trophy. Darwin's insight, inspired by Malthus, was to see that the struggle is actually between different individual members of the *same* species, competing for resources in the same ecological niche.

The correct analogy is with individual football players at a club, competing with one another for a place on the first team. The hawk swooping on a rabbit to obtain its next meal is not competing with a swallow (or even, indeed, with the rabbit); it is competing with *other hawks*, and the individuals that are more successful (which boils down to finding food and getting a mate) will be the ones that reproduce and pass on their characteristics to the next generation. The only gap in Darwin's picture of this struggle for survival was that he did not know how the characteristics of an individual are passed on to succeeding generations – but that mystery would not be solved until the 20th century.

According to Darwin's *Autobiography*, it was 'in June 1842 I first allowed myself the satisfaction of writing a very brief abstract of my theory in pencil in 35 pages'. This outline survives among Darwin's papers, and is indeed dated 'May & June 1842'; but there is also a thirteen-page outline, written in ink and undated, which Darwin does not mention in his *Autobiography* but which several historians persuasively argue dates from 1839, shortly after Darwin became familiar with Malthus' ideas.[4] The pencil sketch is, though, more complete, and shows that Darwin's ideas about evolution were essentially in place by the time that he moved from London to Down House.

Darwin unveiled his ideas cautiously to a few trusted friends and colleagues, including George Waterhouse at the British Museum, and young Joseph Hooker, a botanist who was to become one of Darwin's closest collaborators and supporters. Hooker had been born in 1817, eight years after Darwin, and returned from a voyage as assistant surgeon and naturalist on a ship exploring the southern hemisphere in 1843. He eventually succeeded his father, Sir William Hooker, as Director of the Royal Botanic Gardens at Kew; it was the two Hookers who made Kew (the English equivalent of the Jardin du Roi in Paris, where Buffon had worked) a scientific centre of excellence for botanical studies.

Hooker provided a natural foil to Darwin's theorising, with a shared background of travel and overseas study. Over the years, he slowly and cautiously came to accept the theory of evolution by natural selection, sufficiently doubtful to make Darwin think carefully about how to present his arguments, but sufficiently encouraging to help carry those arguments forward. After their first exchanges of ideas, in the spring of 1844 Darwin went back to his pencil outline of the theory and developed it into an essay running to 189 pages of manuscript, about 50,000 words. He was sufficiently pleased with the result to have a fair copy (which came out at 231 pages) made by the local schoolmaster, and sufficiently sure of its importance to science to write a letter to Emma, to be opened in the event of his death, requesting her to ensure that it was published. But he had no intention of publishing these dangerous ideas while he was still alive.

Darwin's appreciation of the personal problems that publication of his theory would cause was reinforced in October 1844, when a strange book called *Vestiges of the Natural History of Creation* was published.

The author was Robert Chambers, the founder of the publishing house which bears his name. A 'serious' amateur geologist, Chambers did not restrict his book to areas where he was reasonably expert, but wrote (in a highly accessible and entertaining way) about the origins of the planets and of life, the fossil record, and the origins of humankind itself. Although his book went through many editions, and his identity as the author was soon unmasked by the cognoscenti, it never appeared under his own name in his lifetime. Although a mish-mash of half-truths, speculation and a few bits of good science, the underlying theme of the book was decidedly evolutionary, and Chambers, a respectable gentleman with a family of eleven children to provide for, did not intend to let the backlash it provoked damage his status in society.

Hooker rather liked the *Vestiges*, for all its faults; Darwin responded that the author's 'geology strikes me as bad and his zoology far worse'.[5] No doubt he was thinking of the passage in which Chambers suggested that humankind was descended from a large frog. It would not have been a good time for Darwin to go public with his own evolutionary ideas, although he developed them further over the next couple of years, largely on the basis of discussions with Hooker. Intriguingly, though, at the same time he actually did publish most of his ideas about evolution, not explicitly but almost in a coded form.

The vehicle he used was the second edition of his *Journal*, describing the *Beagle* voyage. Darwin worked on this during 1845, when *Vestiges* was at the height of its success, and added bits and pieces of material, a paragraph here and a paragraph there. Taken in isolation, these additions are nothing striking. But a comparison between the first and second editions of the *Journal* makes it possible to take out all the additions and put them together. As Howard Gruber has pointed out, 'taken out of their hiding-places and strung together, they form an essay which gives almost the whole of his thought'.[6] Gruber describes Darwin's ploy as involving two kinds of concealment, one being to break up his ghost essay into pieces scattered throughout the second edition of the *Journal*, and the other being to leave out one vital ingredient, the principle of natural selection itself.

This discovery, together with the letter to Emma about the essay of 1844, shows Darwin's state of mind at the time. He was fired with enthusiasm for his ideas, and wanted the world to know about them; but he was equally concerned at the opprobrium that would inevitably fall not only on his head but on the family as well. He was also aware that he lacked the scientific standing to be taken seriously if he did put forward his theory of evolution. He was, after all, still very much known as a geologist. Why should biologists take his ideas any more seriously than those of the author of the *Vestiges*?

The remedy, as we have seen, lay in immersing himself in the epic study of barnacles. It gave him something to do other than worry about whether or not to publish his ideas on evolution; and it also, as Darwin himself appreciated, gave him the expertise he would need to be taken seriously when he did decide to break cover on the species problem, and the tools with which to defend his theory from the inevitable attacks.

Of course, Darwin kept up his evolutionary studies as well. Barnacles themselves provided more evidence of evolution at work, and he maintained his correspondence with others who were interested in the origin of species. One of these correspondents was Edward Blyth, the curator of the Asiatic Society's museum in Calcutta; another was Alfred Russel Wallace, a naturalist based in the Far East.

Wallace was even younger than Hooker, and had been born in 1823. His background was very different from that of Darwin, with no inherited wealth to rely on, and no illustrious scientific (or poetical) ancestors. He was the eighth of nine children, the son of an unsuccessful solicitor, and had to leave school at thirteen to earn a living. After a year working for a joiner, he became an apprentice to his elder brother William, a land surveyor, and spent the years up to the age of 20 helping to survey railways, canals, and the land enclosures that were a hot political issue at the time. This took him out in the countryside and encouraged him to become a keen naturalist, who made his own observations and read avidly about the natural world.

In 1843 Wallace lost his job as the country was gripped by an economic recession, and he worked for a time as a schoolmaster, continuing to study natural history, and becoming friendly with a colleague, another eager amateur naturalist, Henry Bates. Among the books they read and discussed together were Malthus' *Essay* and the infamous *Vestiges*, which set them thinking about evolution. They also read Lyell's *Principles of Geology* and the new edition of Darwin's *Journal*.

When a renewed attempt at surveying work also failed, in 1847, Wallace decided to try his luck as a freelance naturalist in the New World. This was not a completely crazy scheme; there was a great demand for new specimens of all kinds by both private collectors and museums, and good prices would be paid for rare examples. Together with Bates, Wallace planned an expedition to the Amazon, first boning up on natural history at the British Museum, and seeking help from Sir William Hooker at Kew. They both needed money, they were both keen naturalists, and neither had any family ties. What better way to make their fortunes than by seeking rare specimens in the jungles of South America? They set sail in 1848, when Wallace was 25, three years older than Darwin had been when he departed on the *Beagle*.

More than four years spent exploring and collecting under difficult conditions gave Wallace the same sort of firsthand knowledge of natural history that Darwin had gained some twenty years earlier; but disaster struck on his return voyage. The ship in which he was travelling was lost by fire, taking virtually all of his collection with it, and the crew and passengers were only rescued after ten days at sea in open boats. Wallace returned to England as penniless as when he had left, with no collection to sell, but armed with a much greater understanding of the natural world. Bates, who had stayed in South America, returned three years later, without any significant damage to his specimens.

Undaunted by the disaster, Wallace stayed in England only long enough to regroup and plan another expedition. He presented papers to several scientific societies and published his *Narrative of Travels in the Amazon and Rio Negro* during the fifteen months he was in England, and then set off once again, this time to South-East Asia. The reputation he established with his scientific contributions enabled him to get a free passage on a navy ship early in 1854; Darwin, who was nearing the end of his epic study of barnacles, read some of Wallace's published work, and was impressed immediately by what the 30-year-old self-taught naturalist had to say about the extreme variability of species of butterfly in the Amazon valley. This led to an exchange of letters between Darwin and Wallace that continued intermittently over the next few years, while Wallace was naturalising in and around Borneo. It also led to Darwin becoming one of Wallace's customers, purchasing specimens from him, and sometimes complaining mildly in his private notes about the cost of having them shipped back to Britain.

He was eager for more specimens because, with the barnacles out of the way, Darwin's thoughts were concentrating once again on the species question. 'From September 1854 onwards,' he wrote in his *Autobiography*, 'I devoted all my time to arranging my huge pile of notes, to observing, and experimenting, in relation to the transmutation of species.' But *why* should species 'transmute' – diverge from a common stock? By now, Darwin thought he had the solution. He tells us that the moment the idea struck him remained clear in his mind, so that 'I can remember the very spot in the road, whilst in my carriage, when to my joy the solution occurred to me'. Unfortunately, he doesn't tell us either where the spot was, or the exact moment that the inspiration struck, only that 'this was long after I had come to Down'. Darwin's solution to the problem of why speciation occurs was that 'the modified offspring of all dominant and increasing forms tend to become adapted to many and highly developed places in the economy of nature'.

This was very much an idea of its time, in the first industrialised economy in the world. Division of labour and specialisation in the industrial world were familiar concepts to Darwin, both in the abstract and by his association, through blood and marriage, with the Wedgwood clan, whose fortunes were founded on the successful application of production-line factory techniques to their business. Any individual that could exploit a vacant niche in nature would be successful, Darwin realised, like an entrepreneur who identified a gap in the economic market place and moved in to it.

He was still reluctant to publish, but in April 1856 he at last confided in his mentor Charles Lyell, when he visited Down House for a few days. This was no impulsive decision, but part of a careful broadening of the circle of colleagues with whom he could discuss evolution. In the same month, Darwin also sent invitations to Hooker, Huxley and Vernon Wollaston, a beetle expert from the British Museum, to attend a weekend gathering at Down House to thrash out ideas. Also invited, but unable to attend, was Hewett Watson, a botanist who had expressed evolutionary views and in whom Darwin had already partially confided. Lyell was informed of their discussions, and of the extent of Darwin's theory.

Astonished, and far from being convinced entirely, Lyell wrote to his wife's brother-in-law, the botanist Charles Bunbury, 'I cannot easily see how they can go so far, and not embrace the whole Lamarckian doctrine.'[7] But of one thing he was certain – whether or not he could go the whole hog with Darwin, these ideas were too important to be restricted to discussions among a narrow circle of scientists meeting for the odd weekend at a country house in Kent. On 1 May, Lyell wrote to Darwin, urging him, in no uncertain terms, to publish:

I wish you would publish some small fragment of your data, *pigeons* if you please and so out with the theory and let it take date and be cited and understood.[8]

Like most modern scientists, Lyell was worried about establishing priority, the idea of publishing just a brief paper so that Darwin could prove, if the theory did turn out to be correct, that he had come up with the idea before anyone else had. That was far from Darwin's mind, partly because he genuinely was not bothered greatly about establishing his priority to the idea, and partly because he seems to have had a blind spot concerning the possibility of someone else coming up with the idea. He had sat on it now for nearly twenty years, and it seems never to have occurred to him that in all that time someone else might follow the same path that he had to arrive at the same conclusions. Nevertheless, the outcome of these meetings in April 1856, when he had at least expanded slightly the audience for those ideas, was that, prompted by Lyell and encouraged by Hooker, he at last decided that he would publish – not a brief paper that could hardly do justice to the theory, but a proper book, a weighty scientific volume that might take him two years to complete. And he also spread his ideas still further afield, writing about them in a letter to Asa Gray, an American botanist with whom he corresponded, which provided just the kind of outline of the entire theory that Lyell had urged him to publish.

One reason why Lyell was so concerned about Darwin establishing his priority to the theory of evolution was that he had read a paper from Wallace published in September 1855, in the *Annals and Magazine of Natural History*. The paper reviewed the evidence for evolution, under the title 'On the Law which has Regulated the Introduction of New Species', but did not offer a theory of how and why evolution might have occurred. Darwin seems to have missed the paper initially, but it was pointed out to him both by Lyell and by Blyth, who wrote exuberantly 'what do you think of Wallace's paper in the *Ann. M. N. H.*? Good! Upon the whole.'[9]

The publication of Wallace's paper, and Blyth's response to it, shows how much the climate of opinion had shifted over the two decades that Darwin had been hugging his theory to himself. What was unthinkable at the end of the 1830s was at least a subject of open discussion by the middle of the 1850s, and the word 'evolution' could be aired in respectable scientific circles, as well as in the pages of the *Vestiges*. But, perhaps because of his isolation in Down House and his reluctance to attend scientific meetings, Darwin still seemed to have been blind to the possibility that Wallace was on the same trail that he had followed, although he did take the trouble to send Wallace a coded 'hands off' message in a letter offering his congratulations on the *Annals* paper. In May 1857, hoping to make his own position clear without giving his hand away, as well as commenting that 'we have thought much alike and to a certain extent have come to similar conclusions', he wrote:

This summer will make the 20th year (!) since I opened my first note-book, on the question how and in what way do species and varieties differ from each other. I am now preparing my work for publication, but I find the subject so very large, that though I have written many chapters, I do not suppose I shall go to press for two years.[10]

Judging by the time it had taken to complete the barnacles study, this was a wildly optimistic assessment of when Darwin might be ready to publish, not least since he was guilty of a little white lie in referring to the 'many chapters' that he had already written. In fact, by the end of 1856, he had finished only three chapters of the intended *Natural Selection* book although he had, of course, already gathered the material for other chapters in rough form.

If the letter of May 1857 was intended as a 'hands off' warning to Wallace, it did not succeed. Indeed, it had exactly the opposite effect. Cut off in the East Indies, Wallace was unaware that anyone had noticed his paper, and wrote back to Darwin expressing his disappointment at the lack of any other response, but indicating his intention of pressing on with his ideas. Responding in December 1857, Darwin stressed both his enthusiasm for Wallace's work and the high opinion his colleagues had of it. 'I am extremely glad to hear that you are

attending to distribution in accordance with theoretical ideas,'
he said, and as for Wallace's paper, 'two very good men, Sir C.
Lyell and Mr E. Blyth of Calcutta, specially called my attention
to it'. All of this rekindled Wallace's enthusiasm, encouraging
him to press on with his theorising. But the blinkers were still
on at Down, as Darwin continued 'though agreeing with you
on your conclusions in that paper, I believe I go much further
than you'.[11]

He was wrong. Wallace's letter to Darwin that arrived on 18
June 1858, containing the manuscript of a paper headed 'On the
Tendency of Varieties to Depart Indefinitely from the Original
Type', asking Darwin to show it to Lyell and requesting their
views on its contents, is often presented in the Darwin story
as a bolt from the blue, a fully fledged theory of evolution
coming from an obscure and unfamiliar naturalist. In fact,
Wallace was known to the scientific community in general,
and very well known (at least as a correspondent) to Darwin,
who had actively encouraged him to develop his theory. The
pedigree for Wallace's version of the theory was, indeed, at least
as good as the pedigree Darwin's theory would have had if he
had published in 1844, when he would have been the same age
(35) that Wallace was in 1858.

But it was too late to think about what might have been in
1844. After 20 years' thinking about evolution, Darwin had been
pre-empted. Wallace might have sent the paper direct to one of
the learned journals, with consequences we can only guess at;
what actually happened is that Darwin duly sent the paper on
to Lyell, with an anguished covering letter.

Your words have come true with a vengeance – that I
should be forestalled . . . I never saw a more striking
coincidence; if Wallace had my MS. sketch written out
in 1842, he could not have made a better short abstract!
. . . [He] does not say he wishes me to publish [his paper],
but I shall, of course, at once write and offer to send it to
any journal. So all my originality, whatever it may amount
to, will be smashed, though my book, if it will ever have
any value, will not be deteriorated; as all the labour consists
in the application of the theory.[12]

Lyell, in consultation with Hooker, came up with a happy compromise. A meeting of the Linnean Society was due to be held on 1 July. Brushing aside Darwin's doubts that it might not be quite the gentlemanly thing to do, they arranged for a joint presentation at that meeting of Wallace's paper, Darwin's 1844 outline of his theory, and the letter to Asa Gray of 5 September 1857, which helped to establish Darwin's prior claim. Darwin himself could not attend the meeting. It is unlikely that he would have done so anyway, but the death of the baby on 28 June, followed by his burial and the evacuation of the other children to stay with Emma's sister Elizabeth on 2 July, made any trips to London out of the question.

The presentation of the evolution theory at the meeting on 1 July caused only a minor stir. The cat was out of the bag at last, but nobody seemed unduly bothered. It was the last meeting before the summer break, six papers in all were presented to the meeting, and there was important Society business to attend to. Even so, it comes as something of a surprise to a modern reader to find that almost a year later, on 24 May 1859 (at a meeting held to mark the anniversary of the birth of Linnaeus), the President of the Linnean Society summed up the events of the past twelve months with the comment:

The year which has passed ... has not, indeed, been marked by any of those striking discoveries which at once revolutionise, so to speak, the department of science on which they bear.[13]

Before the year was out, he must have wished he had never made that remark.

Even when the 'joint paper' by Wallace and Darwin was published in the *Proceedings* of the Linnean Society,[14] it received only a muted, and largely negative, response. The reaction of Samuel Houghton, addressing the Geological Society of Dublin in February 1859, is not untypical. He suggested that the only reason anybody had taken any notice of the joint paper was because of:

the weight of authority of the names [Lyell and Hooker]

under whose auspices it has been brought forward. If it means what it says, it is a truism; if it means anything more, it is contrary to fact.[15]

But by the time those words were spoken Darwin's masterwork was almost ready for publication.

Prompted by the fear of being further pre-empted by a book from Wallace (in fact, Wallace gave up any plans to write such a book after the publication of the *Origin*), in the summer of 1858 Darwin began serious work on what he intended to be an 'abstract' of his great book on *Natural Selection*. At first he intended this to be a substantial scientific paper; by the autumn he had realised that he would need a small book to contain everything he had to say. In the end it turned out to be quite a large book, running to more than 150,000 words.

Even when Darwin began discussing the book with its eventual publisher, John Murray, his working title was still *An Abstract of an Essay on the Origin of Species and Varieties through Natural Selection*; at Murray's prompting, by the time it was published in November 1859 the title had been slimmed down to *On the Origin of Species*, with the words *by Means of Natural Selection* in smaller type, followed by a typical Darwin afterthought *or the Preservation of Favoured Races in the Struggle for Life*. No matter how Darwin might have wished to explain everything in the title alone, though, it has always been known simply as the *Origin*.

Nobody involved expected the book to repeat the commercial success of *Vestiges*. Murray initially planned to print 500 copies, but increased this to 1,250 once he saw the finished work. In the event, the booksellers subscribed for 1,500 copies (so although it was not sold out on the day of publication, as folklore claims, every copy was in the shops) and an immediate reprint (actually a second edition, since Darwin was still tinkering with the text) was set in motion.

We shall look at the public, and scientific, response to the *Origin* in the next chapter; but first, we should spell out exactly what Darwin's theory is all about.

The theory of evolution contains two key ingredients, variation and selection. First, individuals in one generation reproduce to produce individuals in the next generation that are not exact copies of their parents. There is a variety of slightly different individuals, variations on a basic theme, in every generation. Second, natural selection acts on this variety of individuals, winnowing out those which are less well fitted to their ecological niches and leaving only the fittest, in this sense of the word, to survive to reproduce in their turn.* In subsequent generations, individuals will inherit many of the characteristics of those survivors, so they will be well fitted for the life they lead; but there will still be variations on the theme, allowing room for further improvement as Malthusian pressures continue to cause a struggle for survival.

If a longer beak helps a bird to survive, for example, then it will pass on the characteristic longer beak to its offspring. Starting out from a population of birds with a variety of beaks, the average beak length in the next generation will be greater, because birds with shorter beaks will have missed out in the struggle and failed to find food, so they will not leave descendants – remember that the competition is with *other members of the same species*, which all need the same kind of food. In each subsequent generation, as longer beaks are favoured the birds with shorter than average beaks will get less food and reproduce less successfully. It does not matter that even the birds with shorter than average beaks now have longer beaks than their ancestors did. They are not competing with their ancestors, but with their rivals in the present generation, who have longer beaks still!

Darwin's unique contribution was not the idea of evolution, which was being widely discussed in the middle of the nineteenth century. What he and Wallace provided was an explanation of *how* evolution occurs. He did not know how characteristics are passed on from one generation to the next, nor did he know why there was variability among individual members of a population (we shall explain these points in later chapters). But this did not matter, because his observations over many years and his

---

* It is worth emphasising that Darwinian 'fitness' is not necessarily the same as physical fitness. Being big and strong may or may not be an advantage to an individual; what matters is how well the individual fits in to the life-style of its species.

correspondence with other naturalists, farmers, and anyone else who might have something to say on the subject had shown him that this is indeed what happens. He also realised that evolution acts in two ways. First, it keeps an existing species exquisitely tailored to its existing ecological niche, as long as that niche still exists. Some humming-birds have been so 'fine tuned' in this way by evolution that their beaks can only be used to sip nectar from one species of flower – which is fine as long as those flowers are available, but means that the fate of those birds is sealed should the flowers, for any reason, die out.

The second way in which evolution works – and this is really what interested Darwin most – is to create new species from the varieties found among existing species. This may happen when groups of individuals become separated from one another and take up slightly different life-styles (through chance or necessity). The classic example is provided by the birds of the Galápagos Islands, where a few original settlers from the mainland, members of the same species (the finch), ended up on different islands and adopted different life-styles. In one case, evolution favoured the longer beak; on another island, a short, heavy beak was more effective at cracking seeds. Time (many generations) and Malthusian pressures were all that was then needed to produce two different species of bird. Given even more time, the same process, Darwin argued, could explain the evolution of all species of life on Earth from a distant common ancestor.

The *Origin* is almost unique among great scientific works in presenting a revolutionary new theory in language that can be not only understood, but read for pleasure by people who have no background knowledge of science whatsoever. This is certainly not true of, say, Albert Einstein's theory of relativity, or Isaac Newton's *Principia*, two works which are among the few that rank in importance with the *Origin*. So, rather than attempting to paraphrase Darwin further, we would like to give you a flavour of what he actually said in his great book. The whole thing is well worth reading, even today – but if you do seek it out, be sure to read the first edition, which presents Darwin's ideas with the greatest clarity and force. In later editions he made many changes in response to various criticisms, and in almost all cases it has turned out that Darwin was right in the first place, and his critics were wrong. So the *worst* edition to follow is the sixth, the last one which Darwin worked on, even though, ironically, for that reason it is the edition that has been most commonly reprinted. All quotes below are from the first edition.[16]

Early in the book, Darwin gets in a dig at the pseudo-scientists who had embraced a form of evolution:

> The author of the 'Vestiges of Creation' would, I presume, say that, after a certain unknown number of generations, some bird had given birth to a woodpecker, and some plant to the misseltoe, and that these had been produced perfect as we now see them; but this assumption seems to me to be no explanation, for it leaves the case of the coadaptations of organic beings to each other and to their physical conditions of life, untouched and unexplained.
>
> It is, therefore, of the highest importance to gain a clear insight into the means of modification and coadaptation.

That clear insight is just what he goes on to provide. He discusses variation under domestication, including his own studies of pigeons, and then moves gently on to variation under nature, drawing upon everyday examples from the countryside of nineteenth-century Britain and more exotic studies, including the birds of the Galápagos, to establish that the range of variability of a species often overlaps with the definition of different species. 'Amongst organic beings in a state of nature there is some individual variability,' says Darwin, and:

> It is immaterial for us whether a multitude of doubtful forms be called species or subspecies or varieties ... But the mere existence of individual variability and of some few well-marked varieties, though necessary as the foundation for the work, helps us but little in understanding how species arise in nature. How have all those exquisite adaptations of one part of the organisation to another part, and to the conditions of life, and of one distinct organic being to another being, been perfected?

The answer, of course, lies in 'the struggle for life'.

> Natural Selection, as we shall hereafter see, is a power
> incessantly ready for action, and is as immeasurably superior
> to man's feeble efforts, as the works of Nature are to those
> of Art . . .
> I use the term Struggle for Existence in a large and
> metaphorical sense, including dependence of one being on
> another, and including (which is more important) not only
> the life of the individual, but success in leaving progeny.

This is of key importance. In evolutionary terms, the *only* thing
that matters is reproduction. An individual that survives long
enough to reproduce and leave offspring that reproduce in their
turn is, at least to some extent, a success; an individual that leaves
no offspring is an evolutionary failure.

> A struggle for existence inevitably follows from the high rate
> at which all organic beings tend to increase. Every being,
> which during its natural lifetime produces several eggs or
> seeds, must suffer destruction during some period of its
> life, and during some season or occasional year, otherwise,
> on the principle of geometrical increase, its numbers would
> quickly become so inordinately great that no country could
> support the product . . . it is the doctrine of Malthus.

'In looking at Nature,' he says:

> it is most necessary to keep the foregoing considerations
> always in mind – never to forget that every single organic
> being around us may be said to be striving to the utmost
> to increase in numbers.

Darwin describes natural selection as the 'preservation of
favourable variations and the rejection of injurious variations',
and points out that 'as all the inhabitants of each country are
struggling together with nicely balanced forces, extremely slight
modifications in the structure or habits of one inhabitant would
often give it an advantage'. The result is that:

natural selection is daily and hourly scrutinising, throughout the world, every variation, even the slightest; rejecting that which is bad, preserving and adding up all that is good ... we see nothing of these slow changes in progress, until the hand of time has marked the long lapse of ages, and then so imperfect is our view into long past geological ages, that we only see that the forms of life are now different from what they formerly were.

This almost poetical vision of the extent of past ages was to cause problems for Darwin throughout the rest of his life, and led to many of the incorrect modifications of later editions of his masterwork, for reasons that we discuss in Chapter 12. The aspect of the *Origin* that would cause intense excitement at the end of the 1850s, however, was the inevitable implication that people had evolved in the way that Darwin described. He never said so explicitly in the book, but it was easy to read between the lines. Mentioning the idea of life being represented not by a ladder of evolution but by the growing tips of the twigs of a great tree, he said:

I believe this simile largely speaks the truth. The green and budding twigs may represent existing species; and those produced during each former year may represent the long succession of extinct species. At each period of growth all the growing twigs have tried to branch out on all sides, and to overtop and kill the surrounding twigs and branches, in the same manner as species and groups of species have tried to overmaster other species in the great battle for life.

Still concerned by the time-scales required for all this, Darwin stresses that 'the lapse of time has been so great as to be utterly inappreciable by the human intellect':

If we admit that the geological record is imperfect in an extreme degree, then such facts as the record gives, support the theory of descent with modification. New species have come on the stage slowly and at successive intervals; and the amount of change, after equal intervals of time, is widely different in different groups ... The gradual diffusion of dominant forms, with the slow modification of their descendants, causes the forms of life, after long intervals of time, to appear as if they had changed simultaneously throughout the world. The fact of the fossil remains of each formation being in some degree intermediate in character between the fossils in the formations above and below, is simply explained by their intermediate position in the chain of descent.

In a mighty blast at the Creationists, Darwin prophetically writes, referring to the notion of a series of special creations, that:

The day will come when this will be given as a curious illustration of the blindness of preconceived opinion. These authors seem no more startled at a miraculous act of creation than at an ordinary birth. But do they really believe that at innumerable periods in the earth's history certain elemental atoms have been commanded suddenly to flash into living tissues? Do they believe that at each supposed act of creation one individual or many were produced? Were all the infinitely numerous kinds of animals and plants created as eggs or seed, or as full grown? and in the case of mammals, were they created bearing the false marks of nourishment from the mother's womb?

Warming to his theme, Darwin at first cautiously suggests that 'animals have descended from at most four or five progenitors, and plants from an equal or lesser number'. But immediately he throws all caution to the winds:

Analogy would lead me one step further, namely, to the belief that all animals and plants have descended from some one prototype ... probably all the organic beings which have ever lived on this earth have descended from some one primordial form, into which life was first breathed.

All organic beings. That must include human beings. But now Darwin decides that enough is enough. He will save that for later. Near the end of the *Origin*, he simply comments, in one of the most famous phrases in science, that 'light will be thrown on the origin of man and his history'. Darwin himself eventually shed some of that light in his later book *The Descent of Man*.

But there was no way that the implications for human evolution were going to be overlooked. It was the realisation that if Darwin was correct then humankind was itself just one animal species among many that kept Lyell, for many years, from giving his full endorsement to the theory. And it was the possibility, in the language of the day, that 'man' was descended 'from' the apes that provided the immediate focus for the debate about Darwin's theory of evolution as soon as the *Origin* was published.

Notes

1, 4, 5, 9, 10, 12, 15 Ronald Clark, *The Survival of Charles Darwin*, Weidenfeld & Nicolson, London, 1984.

2 Quoted by, for example, Antony Flew, *Malthus*, Pelican, London, 1970.

3 See 'Darwin's Notebooks on Transmutation of Species', *Bulletin of the British Museum (Natural History), Historical Series*, 2 (1960).

6 Howard Gruber, *Darwin on Man*, Wildwood House, London, 1974.

7 Quoted by Adrian Desmond and James Moore, *Darwin*, Michael Joseph, London, 1991.

8, 11 Quoted by John Bowlby, *Charles Darwin: A New Biography*, Hutchinson, London, and Norton, New York, 1990.

13 *Journal of the Proceedings of the Linnean Society, Zoology* IV (1860), p. viii.

14 *Journal of the Proceedings of the Linnean Society*, Zoology II (1858), p. 45.

16 Charles Darwin, *On the Origin of Species by Means of Natural Selection, or the Preservation of Favoured Races in the Struggle for Life*, John Murray, London, 1859 [one of the relatively few modern reprints of this edition is published by Penguin].

Chapter 11

Battles with Bigotry

Since I read Von Bar's essays, nine years ago, no work of Natural History Science I have met with has made so great an impression upon me, and I do most heartily thank you for the great store of new views you have given me.
(T. H. Huxley writing to Darwin after receiving his review copy of the *Origin*, 23 November 1859)

If transmutations were rapidly occurring ... the favourable varieties of turnips are tending to become men.
(Samuel Wilberforce in the *Quarterly Review*)[1]

I have read your book with more pain than pleasure. Parts of it I admired greatly, parts I laughed at till my sides were almost sore; other parts I read with absolute sorrow, because I think them utterly false and grievously mischievous. You have *deserted* ... the true method of induction, and started us in machinery as wild, I think, as Bishop Wilkin's locomotive that was to sail with us to the moon.
(Letter from the Reverend Adam Sedgwick to Darwin after reading the *Origin*)

For myself I really think it is the most interesting book I ever read, and can only compare it to the first knowledge of chemistry, getting into a new world or rather behind the scenes.
(Erasmus Darwin writing to his brother after receiving his copy of the *Origin*)

Not impressive from want of luminous and orderly presentation.
(George Eliot)[2]

As can be seen from this mixed bag of responses to Charles
Darwin's most famous work, it was with the publication of the
Origin of Species that the battle lines over evolution were drawn
up and publicly argued over by the members of the scientific
community and the world at large. Up until then a mere handful
of individuals knew about and discussed the subject and no one
but the members of Darwin's inner sanctum knew anything
about his theory of natural selection. According to both Darwin
and Wallace it was this that lay at the heart of the whole idea,
the mechanism by which species could change over vast time
periods evolving from simpler to more complex forms; and,
because of its implications for the place of Mankind in the
larger scheme of things, it was natural selection that caused the
greatest controversy.

At first glance, the row over evolution would appear to have
started incredibly slowly in that the notions expressed in the
Origin did not filter through to the general public or become
an issue for the national media for some considerable time. But
it is important to place the *Origin* in the context of the 1860s.
Without mass access to information, outside the tiny community
of intellectuals the news spread slowly. Although the highbrow
newspapers reviewed the book and the subject was discussed
in the pages of *The Times*, there was no serialisation, advertising
or glossy magazine interviews with the controversial writer, no
television and radio interviews or book signings.

When the *Origin* appeared in the shops it sold out almost
immediately, and a second print run of 3,000 copies was
ordered to supplement the first batch of 1,250 copies; but
these were bought exclusively by the wealthy elite. Even today
a sale of several thousand hardbacks in Britain is regarded as a
huge success. Taking account of the lower population and the
vastly smaller proportion of educated people affluent enough to
buy books during the 1860s, such sales figures were staggering
and indeed the immediate popularity of the book stunned the
ever-pessimistic Darwin. But, despite the huge interest shown in

the book, awareness of the central tenets of Darwin's work still took time to filter through, and as we shall see later, it was not directly through people reading the *Origin* that the word spread. Instead it was through a spin-off of Darwin's achievement – the unequivocal and uncompromising evangelism of Thomas Huxley.

The first serious and well-publicised confrontation (at least within the educated middle-class establishment) was the verbal battle between Darwin's supporters and his disclaimers at the now-famous meeting of the British Association for the Advancement of Science (BAAS) held in Oxford on Saturday, 30 June 1860, some six months after publication of the first edition of the *Origin*. Darwin was naturally absent but his 'generals' Hooker and most especially Huxley, having been well trained and prepared for years to do for evolution what Darwin could never manage – to fight for its validity against the bigots and disclaimers – stood for the cause, while the author pottered around Down House.

It has become fashionable recently to underplay the events of that evening,[3] but there is no escaping the fact that the bitter verbal exchange between the Bishop of Oxford, Samuel Wilberforce, and Huxley could be seen as the first time that the Church and the scientific establishment clashed publicly. Thanks to the vividness of the conflict expressed in the larger-than-life personalities involved, it generated the interest of the newspapers and spread far and wide the notion that the unsteady truce drawn between theology and science was about to be demolished.

As already mentioned, the pressure had been building for some time. Darwin had been terrified that when his work did finally go public he would be castigated by almost everyone. He was, of course, overanxious about the whole thing and this caused him very real physical and psychological distress. But by the time the *Origin* appeared, the intellectual *Zeitgeist* had changed in Darwin's favour. Although almost no one knew of natural selection, the difficulties in allowing for the existence of traditional theology in the face of a ceaseless onslaught from science were becoming insurmountable. Wilberforce was fully aware of the conflict and had already given Darwin a somewhat less than flattering review of the *Origin* in the ultra-conventional *Quarterly Review*. Before the Oxford meeting, he had been tutored in the latest concepts of the evolutionists by his friend and scientific touchstone Owen and he was a man of undoubted oratorical ability and forcefulness of character. At the meeting, Wilberforce was undone by a combination of overconfidence in both his own abilities and his argument, and had come up against a man who was at least as passionate about the science–religion dichotomy as he himself was. Huxley had been spoiling for a fight for years and had already lambasted Owen and Wilberforce in his lectures at the School of Mines and in his publications. That evening, before an audience of some seven hundred in the library of the newly built Oxford University Museum, Huxley had his chance and, with devastating effect, he took it.

The conflict almost never occurred. Uninspired by the evening's programme, earlier that day Huxley had decided to return to London. Then, taking a walk around Oxford, he bumped into the author of *Vestiges*, Robert Chambers. Chambers hated Wilberforce and his cronies not least because over a decade earlier he had been humiliated in a full-blown verbal battle with Wilberforce at a BAAS meeting in 1847. He knew that Wilberforce would be attending that evening's talk – a paper delivered by a little-known American on the application of Darwin's theory to the development of society – an area of intellectual activity that Huxley and Hooker considered quackery. Fortunately Chambers managed to persuade Huxley that he should attend and so events which would cause the first and one of the most famous clashes between Darwinism and the theological establishment were set in motion.

According to eye-witnesses, the evening attracted the largest audience of the Association meeting and the atmosphere in the packed library was electric. After the talk the Bishop asked the chair for permission to counter-argue and spent half an hour denouncing Darwin's ideas, obliquely attacking Huxley and the supporters of evolution and attempting to produce evidence supporting Creationism.

Known as 'Soapy Sam', Wilberforce was a talented speaker and knew how to deliver a talk with great charm and per-suasiveness. Sadly for him, on this occasion, he went too far. Concluding his diatribe, he turned on the evolutionists and said: 'What have they to bring forward? Some rumoured statement about a long-legged sheep.' He then turned to Huxley and said: 'I should like to ask Professor Huxley, who is sitting by me, and is about to tear me to pieces when I have sat down, as to his belief in being descended from an ape. Is it on his grandfather's or his grandmother's side that the ape ancestry comes in?' Huxley was then called to speak and masterfully turned the tables on Wilberforce by replying:

> I should feel it no shame to have risen from such an origin. But I should feel it a shame to have sprung from one who prostituted the gifts of culture and of eloquence to the service of prejudice and of falsehood.

According to some, this conflict between Huxley and Wilberforce has been blown out of proportion and it has been suggested that Wilberforce was joking and trying to defuse a potentially nasty scene. This does not fit the facts. Wilberforce was an extremely belligerent debater. He was absolutely passionate about his beliefs and had no time for the likes of Huxley. It seems beyond reasonable argument that he would have made light of the matter, particularly in this setting.

Wilberforce's attitude was born out of arrogance and a gross misjudgement of the enemy. For so long the theologians had won the day; yet at that point in history it was clear, even to narrow-minded bigots such as Wilberforce, that science was taking many of the garlands and that religion was under threat of being ghettoised by the upstarts wielding the sword of logic and reason. Wilberforce's own ego and anxiety contributed to his undoing that evening. For Huxley's part, he hated the religious establishment and loathed what he perceived as its ignorance and social arrogance. In reality it would have taken him very little effort to say what he said, it simply tripped off the tongue.

There is also some controversy surrounding Hooker's role in the debate and many claim that it was in fact Hooker and not Huxley who stood up to Wilberforce. This is partially true. Hooker did attack Wilberforce with a lengthy and well-structured argument which he recounted to Darwin shortly afterwards in a letter describing the evening's events:

I hit him [Wilberforce] in the wind at the first shot in 10 words taken from his own ugly mouth – and then proceeded to demonstrate in as few more one that he could never have read your book and two that he was absolutely ignorant of the rudiments of Botanical Science ... Sam was shut up – had not one word to say in reply and the meeting was dissolved forthwith leaving you the master of the field after 4 hours battle.[4]

Although Hooker's attack added weight to the Darwinian argument and helped deconstruct Wilberforce's shaky logic, according to some it did not have the emotional impetus nor the depth of feeling of Huxley's response and therefore met its target with the gathered scientists but not with the layman responding to newspaper reports of the events.

News from the 1860 meeting of the BAAS spread far and wide within the intellectual community and while Wilberforce and his supporters retired to lick their wounds, Huxley, Hooker and vicariously, Darwin, were riding the crest of an ideological wave.

Darwin actually cared little about the religious bigots and was far more sensitive about whether or not people interpreted his work correctly. It is perhaps a mark of how far Darwin had changed and developed intellectually and emotionally during his adult life that the man who once feared vilification from his peers now gave little heed to the concerns of the theologians. Clearly Darwin had outgrown what he must have perceived as the notions of religious idealists who were guided solely by emotion and a fatal cocktail of fear and faith. From somewhat unorthodox Christian he had swung through agnosticism (actually a term coined by Huxley to describe his own position some ten years later), atheism and on almost to a precursor of existentialism. In both the microcosm seen through the lens of his microscope and the macro-scale universe beginning to be revealed by the astronomers of the day, he could see only struggle, temporary victory and defeat; lying beneath it all, black and unmoving, there was a complete absence of meaning. It is no exaggeration to say that from his position in the vanguard of discoveries in evolution and natural selection he was wandering into the chasm and that his unvoiced, often unformulated philosophies were not so far from fashionable Paris some eighty years in the future. Perhaps aware of the idea Neitzsche later articulated as 'if you gaze for long into the abyss, the abyss gazes also into you', Darwin often diverted his gaze from his biological black hole to more mundane matters.

It is an interesting facet of Darwin's life and work that he spent more of his time working in backwaters of biology, away from the glamorous, culture-altering areas of science for which he is most famous. As the furore over evolution raged around him, Darwin became fascinated with orchids, an interest which resulted in his book *On the Various Contrivances by which British and Foreign Orchids are Fertilised by Insects*, published in 1862. At the same time he was gathering more data to support his theories for the next big book, *The Variation of Animals and Plants under Domestication* (1868). Throughout this time, Darwin was content to sit on the side-lines and watch the battles, simply guiding his troops. By now Huxley was proselytising the *Origin* as though it were his own book and at times he was overzealous, stepping beyond what Darwin considered to be appropriate for the cause, coming dangerously close to alienating some of the people who would be most useful to them. He and his fellow Darwinians tried to launch several magazines which were devoted to spreading the word about the new sciences, the modern balance between science and society and the key role for the former in benefiting the latter. The first of these was the ailing *Natural History Review* which Huxley, John Lubbock (son of Darwin's neighbour in Downe and an accomplished naturalist who became an active supporter of Darwinism) and others took over but without success. *Natural History Review* turned into *The Reader* which again foundered soon after its launch. They did eventually get the formula right when they launched *Nature* in November 1869.

In the 1860s Huxley and his friends used their periodicals to advertise and promote Darwinian ideas and as a key weapon in the war waged against religion and old-style scientific elitism, but what was vital to the success of their campaign and the eventual general acceptance of Darwin's ideas was not the fortunes of a scientific journal read by a few hundred intellectuals, not even the later, broader success of *Nature*, nor too the books written by Huxley, or even Darwin himself. Instead, evolution went truly public because of Huxley's idea of delivering lectures to 'the working man'.

Huxley started his working-men's lectures at the School of Mines in the spring of 1861 and from the start they were immensely successful. After the professional scientists, the other group who had shown a natural inclination towards evolution and had been attempting for several decades to publicise their own convoluted version of the concept had been the political radicals, particularly what would now be called the far left. Because the concept of evolution could be easily perverted to suit political ambitions of almost any persuasion, it was abused by those who really had little idea of what evolution was all about.

The success of political extremists in conveying to the public any understanding of evolution was extremely limited, but the whole idea of the mysterious notion of 'evolution' was in the air and the mystery associated with it grew more attractive as the word began to appear everywhere, even in the popular press. Huxley was aware of this political aspect of evolution and used it to convey the theory to his uneducated audience, playing on the notion of self-improvement and the idea that even the lowliest can aspire to and achieve greatness in their own lifetime, a sort of bio-communism. Huxley's intentions were to convey the basic principles behind Darwin's theory to a lay audience, comprising bakers, butlers, boiler-makers and every other category of working man interested in self-education. His method was merely a means to an end.

A born teacher, Huxley was immensely popular with the hundreds of working men who attended his lectures. Utilising all the clichés at his disposal, Huxley was superb at couching extremely obscure concepts in simple terms so that a scientifically and often actually illiterate audience could understand him. And it was perfect timing. Fresh from the triumph of the Great Exhibition and inspired by the highly visible power of technology, the layman was as receptive to new ideas as he would ever be. The people of Victorian Britain had only recently been shocked by the first glimpses of gorillas brought back from the jungles of Africa, and talk of the obvious similarity between the look of the human and that of the gorilla abounded. For many, the Church could not match the apparently real miracles science was able to conjure up before their eyes. God was still feared and the churches remained as full as ever, but unquestioning respect for the clergy was slipping, some of the Bible stories were beginning to sound a little unlikely and the principle that Biblical accounts were sometimes allegorical and not to be taken absolutely literally started to gain popularity.

There is no doubt that Huxley's lectures did much to make the public aware of the name Darwin and the concept of evolution. Huxley's description of evolution was simplistic in the extreme. There was no discussion of the ruthlessness of Nature, the cut-throat struggle for survival, but the basic tenets were conveyed and written up by the hack journalists who attended the early talks.

Huxley himself was delighted by the response to his lectures and frequently wrote to an appreciative Darwin telling him of his progress. In one letter he said of his audience: 'By next Friday evening they will all be convinced that they are monkeys.'[5]

In the decade following the publication of the *Origin* a spate of books appeared on the subject of evolution which covered the arguments for and against, the place of man in the scheme of things, the theologians answer to the blasphemy and books applying Darwin's ideas to sociology. Some of these contributed little, others were immensely supportive such as Huxley's own provocatively titled *Man's Place in Nature*, still others ruthlessly attacked Darwin and his concepts.

Man's Place in Nature was the first and the most confrontational of the books by British Darwinians (as we will see, Darwin's German followers were even more evangelical and controversial than Huxley). It was really a distillation of Huxley's lectures and appeared as a single slim volume in December 1862. Priced to appeal to a wide readership, it further consolidated Huxley's portrayal of Darwinism, at once detailed and erudite but at the same time appealing to the general reader.

Two months later, early in 1863, Charles Lyell's *Antiquity of Man* arrived. It was a commercial success; the name of Lyell was already writ large within the scientific establishment and his geological works, written over thirty years earlier, had sealed his reputation and were still perceived as the standard texts on the subject. Sadly, none of this did anything to alter the contents of *Antiquity of Man*. To Darwin, Lyell's book came as a bitter disappointment; it was stodgy, old-fashioned before it appeared, and misguided. Lyell's support for natural selection in written form would have come as a great boost for the evolutionists and Darwin could not help feeling that the elderly Lyell should by then have been ready to take a firm line – almost as if, as a senior member of the scientific community, Lyell had a responsibility to show his hand. Instead Lyell had delivered a book in which his arguments were non-committal over the matter of natural selection and in which he tried desperately to avoid 'the question of Man'. Although he tried to be polite, Darwin's letter to his old mentor after reading *Antiquity of Man* was laced with bitterness: 'I had always thought that your judgement would have been an epoch in the subject. All that is over with me . . .'

For a time, the publication of *Antiquity of Man* cast a shadow over the Darwin–Lyell relationship and although Darwin still greatly admired the elder scientist for his contribution to other areas of science and for his encouragement with the *Origin*, he could never quite forgive him for his inability to overcome his religious beliefs and seemingly misguided commitment to the importance of mankind.

The matter of Lyell's book was compounded by bad timing because during the spring of 1863, as an advance copy of *Antiquity of Man* arrived at Down House, Darwin was just entering a period during which he suffered his worst state of health to date. The illness lasted over a year and resulted in his having to give up work entirely. For much of this time he was bedridden and could not even potter around his garden tending his pollination experiments and cultivating his exotic plants. By September 1863 things had become so bad that Emma even managed to persuade him to return to Dr Gully's in Malvern.

The trip appears to have done little good. Although Emma visited Annie's grave, Charles could not manage it. In some respects it is clear that the trip to Malvern was more for Emma's benefit than for her husband's and was another event which was unintentionally ill-timed. Within two days of Emma's visit to Annie's grave, the Darwins received a letter from a devastated Hooker telling them of the death of his own adored six-year-old daughter Minnie.

In response, Darwin tried to console Hooker and his wife Frances and their pain brought Darwin's own agony flooding back, but, as he reminded himself in his private notebooks, time was a great healer and that his tears had 'lost that unutterable bitterness of former days.'[6]

By the early 1860s Darwin may have been more able to accept the loss of Annie, but his health had not improved in the slightest. For months after his return from Malvern he spent all day every day on the couch in his study hardly able even to stand or to have the newspaper read to him. The best he could manage was to lie with his eyes closed hour upon hour while Emma read him lightweight novels. And so it continued throughout the winter of 1863 and on into the spring of 1864. At last in April the tide began to turn and his health started to pick up, the sickness subsided and he was able to return to his neglected greenhouses and gradually to begin work on *Variation* again. By this time, having been laid low for nearly a year by his latest illness, Darwin had almost detached himself entirely from the arguments growing more vehement over the meaning of his work. He revelled in positive reports from Huxley and grew angry over what he viewed as the mindless ignorance of unqualified commentators entering the fray through the newspapers and ill-informed books, but he never participated in the argument personally. In this respect his health aided him again and acted as buffer against his critics.

November 1864 was a good month for the promulgation of Darwin's theories. On the 3rd, Huxley, Hooker and a small group of Darwinians including Lubbock, the journalist and writer Herbert Spencer and the aggressively atheistic physicist John Tyndall formed themselves into a group called the 'X Club'. Although the club eventually numbered only nine members, all of them were unerring supporters of Darwin's ideas and even more importantly they were all ascending to positions of influence and power. Huxley and Hooker were rising high in the scientific establishment and forming a cabal with the paleontologist Hugh Falconer and another well-known and respected scientist, George Busk. Collectively the members of the X Club wielded a growing influence within the establishment, especially at the Royal Society. Although initially unsuccessful through the magazines they attempted to launch, their early efforts to gain a broader audience for the advancement of their ideas were successful because of their lectures and popular books and they were also able to influence decisions deep within the scientific community itself.

Their first victory came during the first month of the X Club's existence when they overwhelmed the opposition of the older, orthodox element within the Royal Society by forcing through, by ten votes to eight, the awarding to Darwin of one of the society's most prestigious honours, the Copley Medal. Many of the old school, a group of whom had tried to get the medal awarded to Sedgwick, were outraged that it should go to the creator of such an unorthodox theory as evolution by natural selection, but the X Club won the day. Darwin was delighted and it came as a great tonic for him as he dragged himself out of the pit of illness in which he had lain for so long. It also came as a tremendous snub to his critics.

Lyell was pleased for Darwin and had spoken out in his favour at the final meeting to decide upon the awarding of the medal. Despite their disagreements over the issue of 'the dignity of Mankind', Lyell had the greatest respect for Darwin's scientific acumen. In a letter written soon after the award had gone to Darwin, he said:

> Huxley alarmed me by telling me a few days ago that some of the older members of the Council were afraid of crowning anything so unorthodox as the *Origin*. But if they were so, they had the good sense to draw in their horns.[7]

The Darwinians did not have it all their own way. They may have been able to pull strings within scientific circles, but it is a striking fact that Darwin never received a single civil honour in his own country. As early as June 1859, before the *Origin of Species* was published, the newly elected Prime Minister Lord Palmerston, who was a strong supporter of scientific innovation, put forward Darwin's name for inclusion in Queen Victoria's New Year's Honours List. Prince Albert was also behind the idea and suggested a knighthood. Then, with the publication of Darwin's famous book, Wilberforce, who was one of Victoria's ecclesiastical advisers, pointed out that if she were to honour Darwin, it might appear that she had given her approval of the man's theories and in the process she would greatly upset a large number of her subjects, especially the Anglicans. Victoria concurred and Darwin was passed over.

Although these facts were never made public, the truth was suspected by Darwin's supporters and detractors alike. Darwin himself must have heard of what happened and he would have been thrilled but perhaps puzzled by the honour if it had been awarded. Certainly others around him had received similar recognition for considerably less, but, never a man to value or frequently to care about the approval of his peers, the thought of his being considered for a knighthood probably never would have occurred to him. Equally, if he did know that he had been passed over he would also have considered the reasons and found little dishonour in the matter.

The disapproval of his fellow countrymen was largely not shared by the people of other nations.* During his lifetime Darwin received a plethora of awards and honours from abroad, including the 'Pour le Mérite' from Germany (1867), election as a Corresponding Member of the Imperial Academy of Sciences in St Petersburg (1868) and of the French Institute (1878); in Britain he received an honorary degree from Cambridge University in 1877. The Germans in particular were quick to appreciate the value of what Darwin had to say and the *Origin* sold there in large numbers.

One of Darwin's greatest and most vocal supporters was the German zoologist Ernst Haeckel, who did more to spread the faith than anyone besides Huxley. Although his brand of Darwinian evolutionary theory was laced with confusing links with sociology and the development of culture, he was

---

* Although, as we discuss in Chapter 13, there was one exception to this – the awarding of Corresponding Membership of the French Institute, a decoration which met with some controversy.

a good enough zoologist to keep most of his ideas within the bounds of what Darwin himself would have recognised. An obsessive researcher and writer, during the 1860s and 1870s Haeckel produced a collection of enormous books (some of which stretched to over a thousand pages) on Darwinism and its ramifications. These he sent to his guru, Darwin, as soon as they came off the press and of course Darwin was obliged to read as much as possible of the dense German text and to reply with lengthy expositions and analyses. Often Haeckel would go too far (as much as Darwin with his patchy command of German could tell) and Darwin's theories were convoluted to merge with Haeckel's often iconoclastic appraisal of human society and the future of mankind. In this way, the left-wing Haeckel could be perceived as having inadvertently influenced the ideology of fascism. For, although Haeckel died in 1919, some time before Hitler began his rise to power, the embryonic pseudo-science of Social Darwinism could be seen as having its roots in Haeckel's confused fusion of sociology and Darwinism. In the volatile Germany of the early 20th century such blends almost certainly added zest to the early formulations of fascism and extremist politics.

For Darwin, any confusions arising over the meaning of his theories brought him far more anxiety than the trouble-making religious zealots or omission from an Honours List. His own exposition in the *Origin* had been so clear cut and so firmly rooted in detailed experimental data that the adaptation of his purely scientific ideas to sociology or to half-baked pet philosophies upset him. Darwin liked and respected Haeckel and the German was always made welcome at Down House whenever he visited England, but Darwin could not help finding the man overbearing, loud and absurdly devotional. Darwin knew Haeckel was a clever man with a huge surplus of energy, but he found his over-enthusiasm hard to tolerate and often had to put the German scientist right on various points where the theory of evolution had been misinterpreted.

As Darwin struggled to guide some of his more unorthodox supporters away from abusing the concept of evolution by natural selection, he found he had another rebellion on his hands, this time from the very man who had co-discovered the theory – Alfred Wallace.

During the mid-1860s, Spiritualism began to gain popularity in Britain. In middle-class homes all over the country the gullible and the adventurous at heart began to experiment with table-tapping and attempts to contact the departed. Wallace was somehow caught up in all this and, much to the contempt of Huxley and the undisguised dismay of Darwin, he began to try to conjure up a pseudo-scientific explanation of how evolution worked by a mechanism involving 'a superior being'. Darwin of course was finding himself in the absurd position of once more arguing the case against divine interference, but this time with his fellow discoverer of the theory of evolution! Wallace had decided that, rather than following a random, meaningless procession through time, Nature was guided and the process of evolution had a higher meaning, leading to an evolved form of life controlled by spirits from beyond our world.

Although this nonsense of Wallace's caused a rift in his tenuous association with Darwin, hindering perhaps the forward march of evolution, the two men remained friends. Despite the fact that as he grew older, Wallace became even more misguided and for a time was almost ostracised for his unorthodox beliefs, he and Darwin remained in contact; Darwin, who had always been grateful for Wallace's magnanimity over the priority of discovery issue, managed somehow to tolerate Wallace's aberration.

More worryingly, there were many who misinterpreted Darwin's own version of the theory of evolution either intentionally or through total ignorance. The most famous example of the ignorant anti-Darwinian was Wilberforce. His tirades were perceived by many as little more than hot air, and very few scientists and intellectuals took him at all seriously. The real danger to evolution in those early decades after the publication of the *Origin* came not from the likes of Wilberforce but from those who did understand evolution but found themselves fundamentally opposed to Darwin's explanation of how it worked. Chief among these were Richard Owen, George Douglas Campbell, 8th Duke of Argyll, and most significantly, the biologist St George Jackson Mivart.

Darwin and Owen had fallen out long before the publication

of the *Origin* and it is a measure of Darwin's own reputation and standing that the enmity of such an important figure as Owen did him little real damage. Owen was perceived as *the* establishment scientist, a link between the higher reaches of British society (he was a close friend of Prince Albert) and the experimenters and theorists. As already discussed, Owen had his own distinctive ideas about evolution but they were fundamentally at odds with Darwinian natural selection and the friendship between Owen and Darwin, which had always been a shaky affair based on mutual professional assistance during the 1830s and 1840s, could not survive this disparity in the fabric of their work. Owen wrote scathing reviews of Darwin's books from the *Origin* onwards and did his best to deride natural selection in his own lectures, articles and indeed in conversation with the great and the good. Owen once declared that: 'Darwin is just as good a soul as his grandfather, and just as great a goose.'[8]

Needless to say, Darwin's theories were so strong and so well supported that even Owen's considerable influence could do little but keep Darwin off the Honours List and further develop the mystique surrounding Darwinism. Yet Owen did upset Darwin personally and Darwin became convinced that Owen's hatred was not merely an expression of his distaste for natural selection. Writing to Henslow shortly after publication of the *Origin*, Darwin declared:

Owen is indeed very spiteful . . . The Londoners say he is mad with envy because my book has been talked about; what a strange man to be envious of a naturalist like myself, immeasurably his inferior.[9]

Argyll was a successful politician who managed to find time to study and to write highly regarded books on natural history. From the outset he despised the notion of natural selection, probably because he saw in its application the potential for social disruption as well as its threat to orthodox Christianity. In 1867 he published *Reign of Law*, an expertly written and erudite attack on the theme central to the *Origin*. For Argyll, Nature was not a random, chance-driven phenomenon; instead, he believed, evolution occurred by pre-ordained divine design.

His ideas were almost a modern version of Paley mixed with an acceptance of evolution. Naturally, Owen, Lyell and the rest lapped it up. In fact, Argyll had learned his biology directly from Owen, so it was not surprising that *Reign of Law* received glowing reviews from the anti-Darwinians and sold thousands of copies during the 1860s and 1870s.

Although Huxley dismissed the book and its author (calling Argyll the Dukelet), Darwin was worried by it, not least because Argyll had highlighted an apparent weakness in the theory of natural selection – Darwin's inability to explain why Nature developed beautiful-looking creatures such as the humming-bird and the peacock. At the time, explaining this was a problem for Darwin, but it was successfully dealt with soon after in *Descent*, as mentioned in Chapter 8. While Argyll ruffled Darwin's feathers and Owen attempted constantly to undermine the X Club and the growing bastion of Darwinism, it was a former supporter of the theory and peripheral member of the Huxley crowd who caused the greatest disturbance in the rise of Darwin's theory.

St George Mivart was the son of a wealthy hotelier, had been educated at Harrow and was set to enter a career in law when he became fascinated with natural history and zoology after hearing Owen lecturing in London. By 1859, at the age of 32, he was introduced to Huxley and, with the help of references from the unlikely pair of Owen and Huxley, in 1862 he managed to secure a position as lecturer at St Mary's Hospital in Paddington. Shortly afterwards Huxley introduced Mivart to Darwin and although he was never a member of the X Club, Mivart was drawn into the circle of colleagues of whom Huxley, Hooker, Busk and the other Darwinians were the key players. Yet, all the time, as Mivart studied and wrote on zoology, assisted Darwin with experimental material for *Variation* and then on *The Descent of Man*, he continued to harbour deep-rooted, irreconcilable doubts about the mechanism via which evolution worked. Coming from a strict Roman Catholic background, he was trying simultaneously to accommodate a genuine enthusiasm for innovative science and an emotional attachment to his religious faith, eventually finding himself in a similar confusing position to that of Lyell – unable to disentangle the dignity of

Man and the role of God from the pure biological theory of natural selection.

To Darwin's horror and Huxley's unmitigated disgust (Huxley loathed Catholicism long before he knew Mivart), in early 1871, within weeks of Darwin's delivering the proofs of *The Descent of Man*, Mivart published his own *On the Genesis of Species*. With this book Mivart was able to launch the most devastating and thorough attack on Darwinism during Darwin's lifetime. Mivart was from the inner circle; he knew what Darwin was writing about in *Descent* and pre-empted it with his own account but took the opposite tack, returning to the God-controlled evolution espoused by the anti-Darwinians.

Mivart had no personal malice towards Darwin and actually held him in high esteem. Even after the publication of his book, he wanted to continue to engage in intense discussion of the matter of evolution with Darwin, but he was quite naturally perceived as a traitor by Darwin's supporters. The real danger of Mivart's work came not only from the fact that he knew Darwin's mind but, thanks in no small part to the help Huxley had given, Mivart was a highly competent and respected scientist in his own right. Furthermore, he had quite publicly switched sides and provided Darwin's enemies with another thing to crow about.

As we discuss in Chapter 12, Mivart's criticisms of Darwinism were shown to be false eventually and in the long run all his book really managed to do was to slow the process towards an acceptance of Darwinism. Today, while Darwin's *The Descent of Man* is still widely read, *On the Genesis of Species* has been largely forgotten.

While Darwin's work was coming under attack from all sides throughout the 1860s, in many respects his domestic life had reached an enviable dynamic equilibrium. By the time the *Origin* was published, the Darwins had lived at Down House for almost twenty years and their lives had settled into a comfortable routine punctuated only by Charles' irregular but frequent bouts of bad health. By the mid-1860s Darwin could have afforded himself a degree of self-satisfaction with the way his family had grown. After a series of childhood illnesses, Etty's health had settled down and, against Charles' fears, none of his children developed health problems to match his own. Although no great intellect, Etty was a bright girl and indeed, during the winter of 1869, she helped her father with the writing of *The Descent of Man*, demonstrating a definite flair for language. She eventually left home when she married in August 1871.

It had also become clear that Darwin's neuroses about the future careers of his sons were unfounded. The two youngest boys, Leonard and Horace, both attended Clapham School (as did their older brother Francis). Leonard passed the entrance examination to the Royal Military Academy at Woolwich with flying colours, eventually rising to the rank of Major, and Horace followed his older brothers Francis and George to Cambridge University. The eldest, William, meanwhile had gone into a successful career in banking. Quite contrary to Darwin's fears a decade earlier, then at a point where illness looked likely to ravage his family, all the boys enjoyed successful careers: Francis became an accomplished botanist; George a Professor of Astronomy at Cambridge University; and Horace set up his own successful scientific instruments company. After her marriage Etty went on to edit the Darwin family letters. The only child who appears to have been less than averagely able was Elizabeth, known as Bessy within the family. She never married and was content to live at home and to do odd jobs around the house and garden. A quiet and retiring child, she grew into a taciturn and reserved adult about whom very little is known.

If Darwin had sat down to ponder the matter he could not have avoided the conclusion that his life had turned out to be totally different to the way he had imagined it would be twenty years earlier. During the early days of his marriage he had dreaded the world discovering his secret theory and the conflict between this fear and his desire to make his scientific mark on the world caused him ceaseless anguish. In his darkest moments he must have imagined himself becoming a social leper if ever he was to be associated with natural selection. Far from being a social outcast, he was now an esteemed, successful scientist and writer who lived in great comfort in a beautiful house, secure and isolated from much of the flak his theories generated.

On a personal level, the decade was not without its tragedies. In April 1865 Darwin had been shocked by the news of FitzRoy's death. Having led a life constantly on the verge of mental breakdown because of his volatile nature, in a fit of self-loathing sparked off by conflicts with his superiors in the Admiralty, FitzRoy had taken a blade to his throat. Despite his emotional make-up, he had been a remarkably successful man, but he could have made even more of a mark if he had been able to control his temper and his often uncompromising nature. During his career FitzRoy had been an MP, for a short time he had held the post of Governor of New Zealand before being sacked; he was a Fellow of the Royal Society and rose to the rank of Vice-Admiral in the Royal Navy. As a man with strong but often simplistic religious views, he had constantly opposed Darwin's work. He had been at the meeting of the British Association for the Advancement of Science (BAAS) in Oxford that fateful evening during which Huxley and Wilberforce fought out the issue of evolution and he had always been saddened by the fact that in Darwin he had met a man he greatly respected and liked but with whom he would be forever in conflict over fundamental beliefs. Although FitzRoy and Darwin occasionally met and FitzRoy visited Down several times, besides a brief spell at the start of the voyage of the *Beagle*, the two men could never really have been considered friends. They may have come from similar backgrounds, but ethics, religion and science had come between them. 'I never knew in my life so mixed a character,' Darwin said of his old Captain:

> Always much to love and I once loved him sincerely; but so bad a temper and so given to take offence, that I gradually quite lost my love and wished only to keep out of contact with him.[10]

According to at least one commentator on the Darwin–FitzRoy relationship,[11] FitzRoy always blamed himself for introducing what he considered the abomination of Darwin's theory into the world because, if it had not been for him Darwin would never have gathered the experimental evidence for his theory. It was also noted that FitzRoy's religious fundamentalism verged on the insane. Apparently, during the BAAS meeting of 1860 at which he was in attendance, he was seen to stand up at one point during the debate, waving the Bible in the air and yelling 'The Book, The Book'.[12]

Less than a year after FitzRoy's suicide, in January 1866, Darwin's younger sister Catherine died at the age of 54. Married for less than three years, to Charles Langton who had been previously married to Emma's older sister Charlotte Wedgwood, Catherine had been Robert Darwin's favourite daughter and she was viewed by the family as a woman with great but untapped potential. Then, a few months later Darwin lost his second sister in less than a year when Susan died on her 63rd birthday after suffering a long and painful illness. The two deaths sent Darwin into a temporary relapse during 1866 and naturally he could not find the energy to attend either funeral. Susan Darwin was the last member of the family to live at The Mount and Charles' childhood home was auctioned off soon after the funeral.

Erasmus Darwin, himself a year younger than Susan and still in reasonable health, despite leading a life of hedonism, was even more devastated by the loss than was Charles. Susan had been Eras' favourite sister and although he suffered acute pain at her death, he nonetheless took care of all the funeral arrangements and the sale of The Mount.

Shortly after the publication of *Variation* in January 1868, Darwin entered a period of good health, and he made the most of it. He began to socialise in London more frequently, to attend meetings of the Royal Society and to dine with his friends in the capital. He also used the burst of energy accompanying this good spell to start *The Descent of Man*, a work which eventually took almost three years to complete, but which finally sealed his reputation and completed the three-book series on evolution which would stand as his lasting epitaph.

The period at the end of the 1860s, before the publication of the *Descent*, could be viewed as a calm before the storm kicked up by Mivart. It was a time during which Darwin and his supporters could claim a set of victories in the continuing battle over natural selection. In the summer of 1868 Hooker was elected President of the British Association and delivered a highly influential inauguration speech which professed complete adherence to Darwinism and attacked simultaneously the ranks of bigots and ignorant objectors lined up against it. *Variation* was in the shops and despite Darwin's own misgivings about it, it was selling well and further consolidating his position over the question of the mechanism via which evolution worked. Two years earlier Spencer had coined the phrase 'survival of the fittest' to describe the process of natural selection and after initial scepticism for the expression, Darwin himself adopted it and used it in *Variation*. His change of heart had been prompted by the realisation that 'natural selection' could be, and indeed was, being misinterpreted as implying that God controlled the process – the very opposite of what Darwin was trying to convey with the theory.

By the time the *Descent of Man* reached the shops early in 1871, the latest storm over evolution had broken thanks to Mivart's well-judged and knowledgeable full-frontal attack. Although the *Descent* sold incredibly well with some 4,500 in print within two months of publication, many have said that these sales figures were due to Darwin's already established fame.* Also, although Huxley openly attacked Mivart and crushed many of his suggestions with razor-sharp intellectual counter-argument, at the time, some of Mivart's objections to natural selection could not be answered adequately. As we will see in the next chapter, some problems with the explanation of the mechanism of evolution took later scientific discoveries to answer. It is easy to forget that Darwin (and Wallace) were far ahead of their time. Working in an age before there was any real understanding of how characteristics could be passed from one generation to the next, let alone how those characteristics could change and develop over long periods, it is not surprising that Darwin did not have ready answers to absolutely every oddity Nature could reveal. But, once again the tide was turning. Mivart's objections would be the last great attack on Darwinism, the last time that Darwin's theory could not fully explain the details of how evolution worked. Before long, other areas of science had come to the rescue.

---

* This view was encouraged by the fact that, coming as it did in two 450-page volumes retailing for 24 shillings, the book was beyond the pockets of the majority of the general public.

Notes

Unless otherwise specified, quotations in this chapter are from *The Autobiography of Charles Darwin and Selected Letters*, 3 vols, ed. Francis Darwin, John Murray, London, 1887.

1 Paraphrase of Samuel Wilberforce's review of *The Origin of Species* in *Quarterly Review* (Summer 1860).
2 Mentioned in the Introduction to *The Origin of Species*, Penguin Classics, Harmondsworth, 1968, p. 12.
3 J. R. Lucas, 'Wilberforce and Huxley: A Legendary Encounter', *Historical Journal*, 22 (1979), pp. 313–30.
4 J. Hooker to Charles Darwin, 2 July 1860, Darwin Archive, Cambridge University Library, 100: 141–2.
5, 9 *Life and Letters of Thomas Henry Huxley*, 2 vols, ed. L. Huxley, Macmillan, London, 1900.
6 Darwin to Hooker, 1 October, 1863, Darwin Archive, Cambridge University Library, 101: 160/2.
7 C. Lyell to Charles Darwin, Nov 1864. *Life, Letters and Journals of Sir Charles Lyell*, 2 vols, ed. K. Lyell, John Murray, London, 1881, vol. II, p. 384.
8 *The Reign of Law, Autograph Manuscripts of Sir R. Owen*, British Museum, London, Owen Collection, 59.7.
9 *More Letters of Charles Darwin*, 2 vols, ed. Francis Darwin and A. C. Seward, John Murray, London, 1903, vol. I, p. 149.
10 Darwin to Hooker, 4 May, 1865, Darwin Archive, Cambridge University Library, 115: 268.
11, 12 Stephen Jay Gould, *Ever since Darwin*, Burnett Books, London, 1978, p. 33.

Chapter 12

Evolution after the Origin

Darwin left two questions unanswered in the *Origin*. First, he did not try to explain how variability and heredity are linked, so that offspring resemble their parents, but are not perfect copies of either parent. The reason for this omission was simple – he did not know the answer, and although he would spend much of the rest of his life puzzling over the problem, if anything he got further away from the right answer as he grew older. The second omission from the *Origin* was any discussion of the detailed implications of evolution by natural selection for human origins. The reason for that omission was equally simple – Darwin was sure that his book was sensational enough without spelling out the close similarities between ourselves and the other apes. But both omissions were rectified as Darwin returned repeatedly to the theme of evolution in the 1860s and 1870s.

His theory of heredity was first presented to the world at large in 1868, in a self-contained chapter at the end of his book on *Variation*. Tucked away behind a mass of impressive observations about the way the natural world works, it made no dramatic impact; and as Darwin himself acknowledged, other people had been thinking along similar lines, although nobody had developed such a complete model of this kind. It was also, as we have mentioned, wrong; but that in itself is no reason to ignore it, because science progresses by tossing out hypotheses and testing them to see whether they stand up. Only the hypotheses that stand the test of repeated observation and experiment become fully fledged theories that are regarded as revealing some deep truth about nature – the process is, indeed, very much like evolution by natural selection, with survival of the fittest theories. Darwin's ideas are interesting in themselves, as the thoughts of a great man, and also for their historical place in the development (descent?) of ideas about heredity. And they were not so crazy, from the viewpoint of the 1860s.

Darwin's hypothesis was given the name 'pangenesis'. He chose the word 'pan', from the Greek, to incorporate the idea that all of the cells in the body contributed (as we shall see) to the process. 'Genesis', of course, is the part of the name that carries the connotations of reproduction. But before we sketch the outline of the idea, we should highlight one fatal flaw with the whole package, a problem which Darwin himself should have realised was insurmountable.

Darwin, like many of his contemporaries, thought that when two parents combined to produce offspring, then the characteristics of the offspring in some sense represented a blend of the characteristics of the parents. At its simplest, this would mean, for example, that if a tall man and a short woman should have children together, the children would all be of intermediate height by the time they had grown up. This is really bad news for the whole idea of evolution by natural selection, because it would mean that in a very few generations all the interesting variability on which selection can operate would be washed away, and all the individual members of a species would be very similar to one another. It is so completely misguided a notion that we will not elaborate further. The surprise – and an indication of just how far people were groping in the dark at the time – is that Darwin, or any other evolutionist of the nineteenth century, should have taken the idea of blending inheritance seriously at all. Darwin's 'explanation' for the continuation of variability in spite of this blending was that internal and environmental factors stimulated new variations in each generation; but this notion was to lead him ever further up a blind alley.

In the *Variation* Darwin described pangenesis as 'merely a provisional hypothesis or speculation', put forward in the hope that it might stimulate scientific discussion and thereby help scientists to uncover the truth. The essence of the idea is that every cell in the body contributes tiny particles, which Darwin called 'gemmules', which circulate through the body and congregate into the reproductive cells (egg or sperm), from where they are passed on to the next generation. In this way, the characteristics passed on to the next generation, Darwin hoped, might be influenced by what has happened to the individual cells during the course of their life – an element of Lamarckism is built in to the model. If, for example, we imagine that the climate turns colder, then the pangenesis argument would hold that in response to the cold the cells of a living animal would release gemmules which would stimulate the growth of long fur in the next generation.

One of Darwin's main worries about this process seems to have been to convince his readers that entities as tiny as gemmules would have to be nevertheless, they could do the job. 'Their number and minuteness,' he said, 'must be something inconceivable.' But he promptly countered such objections by pointing out the existence of species of fish in which a single individual can produce tens of millions of eggs, and by going on to make persuasive analogies with the known effectiveness of biological material in influencing living things:

> The organic particles with which the wind is tainted over miles of space by certain offensive animals must be infinitely minute and numerous; yet they strongly affect the olfactory nerves. An analogy more appropriate is afforded by the contagious particles of certain diseases, which are so minute that they float in the atmosphere and adhere to smooth paper; yet we know how largely they increase within the human body, and how powerfully they act.

Heredity as an infectious disease! In some ways, Darwin was not so far wrong, after all. He got part of the story backwards; we now know that rather than each living cell contributing a tiny 'gemmule' to the sex cells, it is the repeated division of the single living cell formed by the union of two sex cells that shares out copies of a tiny package of biological material into every living cell. Instead of calling that package of material a 'gemmule', we call it DNA, and every cell in the body of a living creature carries a complete DNA description of the entire structure of the body it inhabits (including the propensity, if any, for long fur). Darwin's comment on the possibility of reproduction depending on the existence of almost 'infinitely minute' pieces of living material, infecting every living cell, was indeed apposite. 'Each living creature,' he wrote, 'must be looked on as a microcosm – a little universe, formed of a host of self-propagating organisms, inconceivably minute and as numerous as the stars in heaven.'

A few years later, commenting on the *Variation*, Darwin wrote:

> Towards the end of the work I give my well-abused hypothesis of Pangenesis. An unverified hypothesis is of little or no value. But if anyone should hereafter be led to make observations by which some such hypothesis could be established, I shall have done good service, as an astonishing number of isolated facts can thus be connected together and rendered intelligible.[1]

It was not to be. Pangenesis never became more than an unjustified hypothesis. But even as an unjustified hypothesis it had some merit, because Darwin was at least trying to find a scientific basis for heredity, a mechanism by which biological information could be passed on from one generation to the next. He was not invoking mystic forces, or taking the soft option of saying that such processes could never be understood by science. As with all of his work on evolution – indeed, as with all of his work – the underlying conviction was that there must be a scientific explanation, one that could be tested by observations and experiments. It was this foundation in the scientific method that made Darwin's whole approach to life, not just his best-known theory, so important. And it was his application of the scientific method that he at last turned to the question of our own ancestry, at the end of the 1860s, after the *Variation* was safely in print.

In 1868 Darwin was 59 years old, and the *Origin* had been in print for nine years (it had, indeed, already been revised several times). In spite of his almost coy, single reference in that book to 'light being shed' on human origins, as we have seen it was the human implications that stirred up the greatest fuss when the *Origin* was published. Darwin was happy to let others, such as Huxley, draw some of the fire of the Church and other critics, and this they continued to do during the 1860s. Huxley published a book on *Zoological Evidences as to Man's Place in Nature* in December 1862, Lyell (although still a reluctant convert to the full implications of evolution by natural selection) published his *Antiquity of Man* early the next year, and in 1864 Wallace wrote a long article for the *Journal of the Anthropological Society* setting out the evidence for a long period of human evolution from earlier forms. They were not alone; the question of human origins and the relationship between people and African apes, stirred by discoveries of fossil remains and stone tools, was widely discussed while Darwin himself was concentrating on the *Variation*.

But still, Darwin had something to say on the subject. The scientific world was, of course, eager for his own views on the application of evolutionary ideas to humankind; and in addition, he went further even than some of his closest friends and colleagues. Wallace, although happy that evolution could explain the existence of every feature of all other species on Earth, and of every feature except one in human beings, stopped short of accepting that natural selection alone could have produced the distinctively large human brain. By 1869 he was arguing that the hand of a greater intelligence was at work, guiding specifically human evolution alone. Lyell also inclined towards the view that a Supreme Being might have rigged evolution in favour of humankind. And these were among Darwin's strongest supporters! So when the *Descent of Man* was published in 1871, while it did not have the impact that it would have had if published alongside the *Origin* in 1859, it was an important and clear statement of the evidence that human beings – every feature of human beings – had evolved in accordance with the same scientific rules that applied to all other forms of life on Earth.

Darwin's own position had long been clear in his own mind, even though, for fear of scandal and through his desire to avoid hurting Emma, he had kept quiet about it for so long. Some of the most intriguing insights into the way his ideas developed have come from studies of the papers he left behind when he died, including a pile of unpublished material labelled 'Old and useless Notes about the moral senses & some metaphysical points'. The notes were written between 1837 and 1840, before Darwin had settled in Down House; far from being useless they presage, in some ways, developments in evolutionary thinking that have begun to be taken seriously only in the past twenty years or so. In a fascinating essay written in 1839, Darwin said:

> Looking at Man, as a Naturalist would at any other Mammiferous animal, it may be concluded that he has parental, conjugal and social instincts, and perhaps others. – The history of every race of man shows this, if we judge him by his habits, as another animal. These instincts consist of a feeling of love <& sympathy> or benevolence to the object in question. Without regarding their origin, we see in other animals they consist in such active sympathy that the individual forgets itself, & aids & defends & acts for others at its own expense.[2]

In this essay, written twenty years before the *Origin* was published, Darwin is already groping towards an understanding of altruism, which has been explained fully in evolutionary terms only within the past few decades.

But the key sentence in this passage, from the point of view of understanding the *Descent*, is the first one. *Looking at Man, as a Naturalist would at any other Mammiferous animal.* That is the heart of Darwin's approach: to treat humankind as just one species among all life on Earth, playing by the same rules and moulded by the same evolutionary forces. The approach is taken up today in the science of sociobiology, which does indeed 'look at Man' as it would 'at any other Mammiferous animal', and which still rouses fierce objections from people unwilling to accept that we have no special place in the evolutionary scheme. So it is hardly surprising that when Darwin finally went public with his views on humankind's place in evolution, in 1871, his book did still cause something of a stir.

In the *Descent,* humankind was indeed discussed 'in the same spirit as a naturalist would any other animal'.

The sole object of this work is to consider, firstly, whether man, like every other species, is descended from some pre-existing form; secondly, the manner of his development; and thirdly, the value of the differences between the so-called races of man.

As usual, Darwin argues his case beautifully and effectively.

He who wishes to decide whether man is the modified descendant of some pre-existing form, would probably first enquire whether man varies, however slightly, in bodily structure and in mental faculties; and if so, whether the variations are transmitted to his offspring in accordance with the laws which prevail with the lower animals.

How can any naturalist, reading those words, fail to nod their assent? The reader is carried through the arguments, step by step, to the conclusion that 'we have given to man a pedigree of prodigious length, but not, it may be said, of noble quality':

Unless we wilfully close our eyes, we may, with our present knowledge, approximately recognise our parentage; nor need we feel ashamed of it. The most humble organism is something higher than the inorganic dust under our feet; and no one with an unbiassed mind can study any living creature, however humble, without being struck with enthusiasm at its marvellous structure and properties.

There speaks the earthworm enthusiast! But just when Darwin had rounded off his work on evolution with this crowning achievement, a thorough and sensible evolutionary view of humankind, the genius stumbled. He had a problem as he already knew, with that 'pedigree of prodigious length'; and now, in the face of the *Descent,* some powerful voices were once again raised against Darwin's ideas.

The idea of evolution itself was by now becoming much too well established to be dismissed; but the idea of natural selection, the blind working of natural processes to weed out the unfit from an array of tiny variations in each generation, looked more vulnerable, and came under attack. A combination of these factors – the time-scale problem and attacks on natural selection – pushed Darwin further and further away from his original position, leading him to embrace ever more closely the Lamarckian idea of inheritance of acquired characteristics, and the notion that changes in the environment would have a direct and immediate effect on individuals, modifying the next generation.

The attacks on natural selection as a mechanism are not worth going into in detail here. Although they caused great concern to Darwin and his supporters at the time, when a full understanding of the hereditary mechanism was developed in the 20th century (as discussed in Chapter 14), it emerged that Darwin had been right all along – or rather, that he had been right in the first place, and should never have changed his mind. But we should just touch on the main alternative put forward, since this was to provide an archetype for other variations on the theme in the decades ahead.

One of Darwin's most vitriolic critics in the 1870s was St George Jackson Mivart, a biologist who managed to combine being a Roman Catholic with being a Fellow of the Royal Society and a Fellow of the Linnean Society. Initially receptive to the ideas outlined in the *Origin*, he could not accept the evidence of human evolution, and responded by attacking the whole basis of natural selection in his book *On the Genesis of Species*, which also appeared in 1871.

Mivart's key argument, which was to be echoed by others later, was that evolution could not proceed by the series of tiny steps that Darwin had suggested, because intermediate forms (such as a giraffe with a neck longer than that of a deer, but not long enough for it to browse on tree tops) would have no advantage. Instead, he argued that evolution occurred when there were major changes in body plan from one generation to the next, between parent and offspring – a deer, in effect, giving birth to a giraffe. The idea has overtones of Special Creation, but it has become familiar in more modern times from a certain kind of science-fiction movie, as a 'mutation'; the process was originally called 'saltation', from the Latin word meaning 'to leap about'. The notion has always had a certain appeal, because it is difficult for human beings, used to life spans measured in a few tens of years, to grasp the way in which an accumulation of tiny changes over a very long period of time really can lead to the evolution of such organs as the giraffe's neck, or an eye.*

Once again, however, when stripped of its emotional over-tones and the suggestion that while all the physical attributes of humankind might have evolved (albeit with the aid of saltations) there was still a requirement for the 'soul' to have a supernatural origin, Mivart's criticism boiled down to a question of time-scales. Any species can be transformed into any other species by a very large number of tiny steps *if there is enough time available*. But how much time – how many generations – had there been for evolution to do its work among life on Earth?

Even in the first edition of the *Descent* itself, Darwin was backtracking. 'I now admit,' he wrote:

> that in the earlier editions of my 'Origin of Species' I probably attributed too much to the action of natural selection or the survival of the fittest. I have altered the fifth edition of the Origin so as to confine my remarks to adaptive changes of structure . . . Some of those who admit the principle of evolution, but reject natural selection, seem to forget, when criticizing my book, that I had the above two objects in view; hence if I have erred in giving to natural selection great power, which I am far from admitting, or in having exaggerated its power, which is in itself probable, I have at least, as I hope, done good service in aiding to overthrow the dogma of separate creations.

In spite of the snarl 'which I am far from admitting', that is almost the writing of Darwin at bay, scurrying around to try to shore up his theory in the face of criticism. More shoring

---

* If you find this hard to accept, the best place to seek enlightenment is in Richard Dawkins' *The Blind Watchmaker*, Longman, London, 1986.

up, in the wake of Mivart's book, led to the sixth and final, and worst, edition of the *Origin*, published in 1872. But Darwin's real problem was not with those who, like Mivart, argued on religious grounds that the theory of evolution by natural selection was flawed. Rather, it lay in a running battle he had had with the astronomers and physicists, in particular William Thomson (who became Lord Kelvin), who had come up with what seemed like incontrovertible proof that the age of the Earth could not be anywhere near long enough to provide the time-scale for evolution that the theory of natural selection required.

The first real scientific attempt to calculate the age of the Earth had been made by Isaac Newton, in his *Principia*, where he estimated that it would take a globe of red-hot iron the size of the Earth 50,000 years to cool down. The argument was simple enough. Obviously, unless something was keeping the Earth hot inside, it *must* be cooling down, and would eventually become too cold to support life; Newton's calculation was a first stab at how long it might be before this happened. It caused no great stir at the time, because in Newton's day it was still quite respectable to envisage a time-scale for life on Earth of no more than a few thousand years.

In the eighteenth century, this line of thought was taken up by the Comte de Buffon, as described in Chapters 2 and 4. Unlike Newton, Buffon actually carried out experiments with balls of iron and other substances, heating them up and then observing how long it took for them to cool down. Using this information, he calculated that starting from a red-hot state the Earth would have taken 36,000 years to cool to the point where life could appear, and a further 39,000 years (making 75,000 years in all) to cool to its present state. But Buffon had not allowed for one important effect – the way in which the cold layer of rock on the outside of the Earth can act as an insulating blanket, holding in heat and slowing the rate at which it escapes into space.

In the early nineteenth century, another Frenchman, Jean Fourier (1768–1830), studied the way heat flows through solid objects, and developed equations to describe this kind of cooling process. In 1820 he wrote down a formula for the age of the Earth, based on these studies. Curiously, as far as anyone knows he never wrote down the number that comes out of that formula. Perhaps he wrote it down and then destroyed the calculation, fearing that he would be ridiculed if he made it public. For, instead of Newton's 50,000 or Buffon's 75,000 years, Fourier's formula gave the age of the Earth as one hundred *million* years. It must have seemed ludicrously long in 1820. But by the second half of the nineteenth century, when these kinds of argument were being taken up by Thomson, even a hundred million years was an embarrassingly short time-scale both for geologists and evolutionists, who required much longer still if they were to explain how the face of the Earth, and the nature of life on Earth, had been shaped by the long, slow uniformitarian processes that we see at work today.

Part of the problem, for Darwin, was that he was up against the acknowledged top man in British science of the time. The other part of the problem was that the arguments against Darwin were right – at least, right within the laws of science as known to Victorian physicists.

William Thomson (1824–1907) graduated from Cambridge in 1845, and became Professor of Natural Philosophy at the University of Glasgow in 1846. He pioneered the development of the science of thermodynamics, describing the behaviour of heat and machinery, which was so important in nineteenth-century industrial Britain, and made a fortune out of his invention of the first practical transatlantic telegraph cable. He was knighted, and then made a Baron (Baron Kelvin of Largs) by Queen Victoria, not for his abstract scientific work, but for his practical contributions to engineering. Today, the scientific temperature scale measuring from the absolute zero of temperature (-273 °C) is called the Kelvin scale, in his honour.

Thomson first commented on the finite age of the Earth in 1852, seven years before the *Origin* was published. He wrote that:

> Within a finite period of time past the earth must have been, and within a finite period of time to come the earth must again be, unfit for the habitation of man as at present constituted, unless operations have been, or are to be performed which are impossible under the laws to

which the known operations going on at the present in the material world are subject.[4]

Soon, however, Thomson realised that the problem did not apply to the Earth alone. By this time, it was clear that the Earth gets most of its heat from the Sun, so that although the fact that the Earth is still hot inside (as shown by the activity of volcanoes) is relevant, we also need to know how long the Sun can have been shining as brightly as it is today. Thomson tackled this problem on and off over the years, and it was also studied at the same time by a German physicist, Hermann von Helmholtz. Both of them realised that the most effective way known to the science of the time to keep the Sun hot involved gravity. Simply by shrinking slowly under its own weight, the Sun could release enough energy to keep it shining as brightly as we see it now for ten or twenty million years – far longer than if it were, say, an enormous lump of coal burning in an atmosphere of pure oxygen.

The first time that this kind of calculation made much impact was in 1862, when Thomson published a paper setting out his reasoning in *Macmillan's Magazine*. By now, the *Origin* had been published, and it was starting to become clear just how long a time-scale Darwin needed for natural selection to do its work. Thomson had discussed the question at the meeting of the British Association for the Advancement of Science in 1861, and again in a paper presented to the Royal Society of Edinburgh in 1862. Darwin had calculated an estimate for the age of the Earth based on the rate at which the chalky rocks of the Weald of Kent must have been eroded, coming up with a figure of 300 million years for this phase of geological activity alone. So there is no doubt who Thomson was sniping at in the *Macmillan's Magazine* article, when he wrote:

It seems, therefore, on the whole most probable that the sun has not illuminated the earth for 100,000,000 years, and almost certain that it has not done so for 500,000,000 years. As for the future, we may say, with equal certainty, that inhabitants of the earth cannot continue to enjoy the light and heat essential to their life, for many million years longer, unless sources now unknown to us are prepared in the great storehouse of creation.

Warming to his theme, Thomson went in for the kill:

What then are we to think of such geological estimates as 3000,000,000 years for the 'denudation of the Weald'? Whether is it more probable that the physical conditions of the sun's matter differ 1,000 times more than dynamics compels us to suppose they differ from those of matter in our laboratories; or that a stormy sea, with possible channel tides of extreme violence, should encroach on a chalk cliff 1,000 times more rapidly than Mr Darwin's estimate of one inch per century?

In fact, Thomson was wrong – not in his calculations, which were impeccable, but in his assumption that there were no forces or sources of energy other than those known to Victorian science. Within 50 years, the discovery of radioactivity showed that there were sources of energy that could keep both the Earth and the Sun hot for long enough to explain the time-scales of geology and of evolution; we now know that the Sun's heat comes from the conversion of hydrogen into helium, the process of nuclear fusion, in which some mass is converted into energy in line with Einstein's famous equation $E = mc^2$. There is enough nuclear 'fuel' available in the heart of the Sun to have kept the Sun hot in this way for about five *billion* years so far, and still to offer the prospect of a further five billion years or so of warmth – a far cry from Thomson's assertion that 'inhabitants of the earth cannot continue to enjoy the light and heat essential to their life, for many million years longer'.

Ironically, however, the reason why Thomson was wrong is precisely the reason that he pointed to, in what was obviously intended as sarcasm. There really are sources of energy unknown to Victorian science 'prepared in the great storehouse of creation', and the physical conditions inside the Sun really do differ sufficiently from those in laboratories on Earth to unlock those storehouses and release the energy they contain.

With hindsight, we can see that Darwin would have been right to stick to his guns, and to insist that if evolution by natural selection required a time-scale of hundreds (or even thousands) of millions of years, then it was up to the physicists to find out how the Sun could have stayed hot for so long. In a sense, by highlighting the inadequacy of the Victorian understanding of the laws of physics, the geological record of the antiquity of the Earth and the theory of evolution by natural selection were together pointing the way towards a new theory of physics, the theory of relativity. Evolution by natural selection *requires* $E = mc^2$, and from one point of view Darwin can be said to have predicted Einstein's great work – Einstein himself was just three years old when Darwin died in 1882.

Except, of course, that Darwin was never so bold as to suggest that the idea of natural selection must be correct, and that therefore the physicists must be wrong. He wrote to Wallace, in 1871, that 'I should rely much on pre-Silurian time; but then comes Sir W Thomson like an odious spectre', and he made repeated attempts to modify his own ideas to match the shortness of Thomson's time-scale. In 1869 Thomson had told the Geological Society of Glasgow that:

> The limitation of geological periods, imposed by physical science, cannot, of course, disprove the hypothesis of transmutation of species; but it does seem sufficient to disprove the doctrine that transmutation has taken place through 'descent with modification by natural selection'.[6]

And although Wallace urged Darwin to stand firm, saying 'I would put the burthen of proof on my opponents',[7] as we have seen, the *Descent* carried the backtracking apology for exaggerating the power of natural selection, and the sixth edition of the *Origin* retreated even further from Darwin's initial position.

So by the time Darwin died, the theory of evolution by natural selection was in retreat. Although the evidence for evolution was widely accepted, nobody in the scientific establishment knew how characteristics were passed on from one generation to the next (pangenesis was not taken seriously even by Darwin's closest colleagues, perhaps no longer even by Darwin himself), and the process by which Darwin himself now envisaged species adapting to their environments was a mish-mash of Lamarckian ideas and the suggestion that changes in the climate, or other external stimuli, directly triggered changes in the organisms themselves.

In many simple accounts of the Darwinian revolution, the impression is given that Darwin produced his theory of evolution by natural selection in 1859, that there was a decade or so of debate involving Huxley and a few other people, and that after the publication of the *Descent* Darwin's theory was well-established and accepted by all right-thinking people. But there are two parts to that theory, the fact of evolution (which was becoming hard to ignore even before Darwin went public) and, much more importantly, the *mechanism* by which Darwin (and Wallace) explained evolution, natural selection.

The reason why Darwin is held in such esteem today is for the theory of natural selection. People outside the world of science often fail to appreciate this. There are still religious extremists to be found, for example, who deny the evidence of evolution, and think that they are thereby attacking Darwin's masterwork; but the idea of evolution itself is not unique to Darwin, and is supported by an overwhelming weight of evidence. On the other hand, there are serious discussions among experts today about the details of how natural selection works, and these debates are sometimes presented in the popular media as attacks on evolution itself. Not at all – such debates are concerned with the finer details of the *mechanism* of evolution, not the question of whether or not evolution has occurred at all.

But it is scarcely any wonder that non-specialists sometimes get confused about the evolution debate. Almost from the moment the *Origin* was published, the strong initial statement of the principle of natural selection was progressively weakened until it had all but vanished in the 1880s. Evolution had been established in the wake of the *Origin*, but natural selection – survival of the fittest – had not. It would not be until many years after Darwin had died that natural selection would be rehabilitated, and that full-strength Darwinism would become the triumphantly accepted standard model of how evolution works. But Darwin could have had no inkling of this during his final years, as he pottered about at Down House, basking in the achievements of his children and continuing his own studies of flowers and earthworms.

Notes

Unless otherwise specified, quotations from Darwin on pangenesis in this chapter from Charles Darwin, *The Variation of Animals and Plants under Domestication*, John Murray, London, 1868.

1 *Autobiography*.
2 The 'Old and Useless Notes' are transcribed by Paul Barrett in Howard Gruber, *Darwin on Man*, Wildwood House, London, 1974.
3 Quotations from Darwin on human origins are taken, unless otherwise stated, from Charles Darwin, *The Descent of Man, and Selection in Relation to Sex*, John Murray, London, 1871.
4 Quotations from William Thomson (Lord Kelvin) are taken from Joe Burchfield, *Lord Kelvin and the Age of the Earth*, Macmillan, London, 1975; see also John Gribbin, *Blinded by the Light*, Corgi, London, 1991.
5, 6 Quoted by Ronald Clark, *The Survival of Charles Darwin*, Weidenfeld & Nicolson, London, 1984.
7 Quoted by John Bowlby, *Charles Darwin: A New Biography*, Hutchinson, London, and Norton, New York, 1990.

Chapter 13

The Final Years

Ever since moving to Down House in September 1842 the Darwins had led their lives in an incredibly organised fashion, following a routine which was hardly ever altered. As the children began to leave home and Charles and Emma grew older, the pattern of their lives became even more mechanical and regulated.

Thanks to Darwin's son Francis we have a vivid and detailed portrayal of his father's everyday life during middle and old age. In his *Reminiscences*, later added to Charles Darwin's short autobiography, Francis Darwin notes that both his earliest recollection of his father and his last memory of him were linked to his father's habit of trudging around the Sandwalk. This is a narrow strip of land near the house which had a tree-lined gravel path around its circumference. Darwin walked this path almost every day and when he first formed the habit, he used to count the number of times he completed the circuit, kicking a flint onto the path at the end of each lap. The Sandwalk was originally rented from Baronet John William Lubbock, the largest land owner in the area and father of John Lubbock (of the X Club), but in 1874 Darwin persuaded the younger Lubbock, who had inherited the estate, to sell him the land.

It was while walking the Sandwalk that Darwin did most of his thinking. Counting the laps and kicking the markers was all part of the mantra guiding the pattern of his thoughts. Summer and winter, unless he was couch-bound by illness, Darwin paced circuit after circuit, grinding away at his theories. It was here that the *Origin* was fine-tuned and the *Descent* conceived, where tears for Annie flowed and where Darwin wrestled with the on-going intellectual dilemmas springing from his controversial ideas. It was also a place where the children played and sometimes joined him on his walks. Only rarely did Darwin insist on solitude, most usually when facing panic over a clash with the establishment or a colleague misinterpreting his work.

What Francis Darwin describes in such detail in his *Reminiscences* is a life of carefully organised routine originally constructed to accommodate Darwin's frequent bouts of poor health — short bursts of intense work sandwiched between plenty of rests and walks with a consistent backdrop of domestic calm and unchanging familiarity kept that way by a devoted Emma:

He rose early, and took a short turn before breakfast, a habit which began when he went for the first time to a water-cure establishment, and was preserved till almost the end of his life. After breakfasting alone at about 7.45, he went to work at once, considering the 1½ hour between 8 and 9.30 one of his best working times. At 9.30 he came into the drawing-room for his letters — rejoicing if the post was a light one and being sometimes much worried if it was not. He would then hear any family letters read aloud as he lay on the sofa.

The reading aloud, which also included part of a novel, lasted till half-past ten, when he went back to work till twelve or a quarter past. He then went out of doors whether it was wet or fine; Polly, his white terrier, went with him in fair weather, but in the rain she refused.

Luncheon at Down came after his mid-day walk. After his lunch he read the newspaper, lying on the sofa in the drawing-room. After he had read his paper, came his time for writing letters. These as well as the MSS of his books were written by him as he sat in a large horse-hair chair by the fire, his paper supported on a board resting on the arms of the chair.

From about half past four to half past five he worked; then he came to the drawing room, and was idle till it was time (about six) to go up for another rest with novel-reading and a cigarette. After dinner he played backgammon with my mother, two games being played each night.* He became much tired in the evenings, especially of late years and left the drawing-room about ten, going to bed at half past ten.

Apart from his final year, when heart disease and old age caused him associated health problems, during the last ten years of his life Darwin was least affected by illnesses related to his weak immune system. During this time he worked furiously, almost as though he were trying to make up for time lost over the years through illness.

Throughout the 1870s Darwin continued to be preoccupied with a number of diverse projects and an ongoing chore was the reworking of his early successful books. Both the *Origin* and the *Descent* went into many editions during his lifetime, and with the *Origin of Species* in particular, Darwin did not simply tinker and update but rewrote entire sections of the book to try to accommodate explanations for a variety of issues which had arisen from the original publication. As we have discussed earlier, sadly, these changes were often for the worse.

In February 1872, with the *Descent* in the shops and selling well, Murray published a cheap version of the *Origin* with a planned cover price of six shillings in order to bring it within the pocket of the general public. Unfortunately, because John Murray reset the book in tiny type to keep down the cost of paper

---

* Both Emma and Charles took these games very seriously and for many years they kept a running total of the score. In a letter to Asa Gray, whose wife when she visited Down was tickled by the rivalry between the otherwise loving couple, Darwin passed on the latest score: 'Emma poor creature has won only 2490 games, whilst I have won, hurrah, hurrah, 2795 games.'[1]

the new version was riddled with errors which then had to be removed at great expense before publication, pushing the price up to 7s. 6d. Even then, during the spring of 1872, sales of the *Origin* jumped from around 60 to over 250 per month.

Darwin was delighted with this success and was particularly pleased that he was reaching a wider audience, realising that, if nothing else, it would provide a further consolidation of his fame ready for the publication of his latest work in progress, *The Expression of the Emotions in Man and Animals*.

Darwin had started making observations and taking notes for the *Expression* in Macaw Cottage soon after his first son William was born in 1839, but had returned to a close study of the nature of animal expression only after the completion and publication of the *Origin* and the *Descent*. Published in the autumn of 1872, *Expression* completed the trilogy of Darwin's evolution books and immediately created a sensation in Victorian Britain. Besides the novelty of the subject – describing and explaining facial and bodily expression from the aggressive stance of a threatened gorilla to the blushes of a Victorian lady – it was one of the first books ever to benefit from the use of photographs. The original version contained seven heliotype plates and although this put up the cover price, *Expression* sold staggeringly well – 5,267 copies on the day of publication.[3]

As usual, while Darwin was writing the *Expression* he continued to work simultaneously on several other more mundane projects. In early 1872 he had become fascinated with worms and conducted experiment after experiment on thousands of the creatures kept in jars in his study and in the greenhouse at Down. As with his earlier obsessions with orchids, beetles, barnacles and other varieties of living thing, Darwin gathered specimens from around the world and he had a simple technique for acquiring them. All it took was a few paragraphs from the famous Mr Darwin to be published in *Nature*, for instance describing the behaviour of the common earthworm found in the garden of Down, for him to be inundated with specimens from readers and fellow naturalists far and wide. His efforts and the help of the readers of *Nature* resulted in Darwin's final book, published in 1881 and entitled *The Formation of Vegetable Mould, through the Action of Worms, with Observations on their Habits*.

During the same decade Darwin pursued other interests. At one time or another the greenhouse was filled with insectivorous plants, orchids and creepers and his experiments resulted in *Insectivorous Plants* (1875), *The Effects of Cross and Self Fertilisation in the Vegetable Kingdom* (1876), a much enlarged edition of *Fertilisation of Orchids* (1877) and *The Different Forms of Flowers on Plants of the Same Species* (1877).

To many, Darwin's life-style was an enviable blend of comfort and intellectual satisfaction. While he continued to lead a quiet life in his country retreat, conducting vast amounts of disparate research and publishing a collection of important and surprisingly popular books, his younger colleagues were finding prestige, success and public responsibility a heavy burden. Both Huxley and Hooker were members of several different government commissions and found themselves required to chair a range of committees and investigative boards. At one point Hooker sat on fifteen different committees at the Royal Society (where he was President from 1873 to 1878) and at the same time fulfilled his duties as Director of the Royal Botanical Gardens at Kew.

Huxley also had his problems. As Secretary of the Royal Society, lecturer still at the School of Mines and newly appointed Lord Rector of Aberdeen University, he was trying to juggle too many responsibilities. Inevitably it took its toll on his health and at the end of 1872 he was diagnosed as suffering from dyspeptic nausea and warned by his doctors that unless he slowed down and agreed to follow a strict diet the condition would become far worse. The illness had been precipitated by overwork, but the final straw came from the stress of a protracted legal battle soon after he and his family had moved house.

The trouble started soon after builders had completed work on the house. The Huxleys' belligerent new neighbour claimed that the construction work had caused damage to his basement and took legal action. By the summer of 1872 the neighbour's claim had failed but Huxley was left with crippling legal fees. Having always been in a financially precarious position, partly through lack of care and partly because he had to support a sister whose husband was an invalid, Huxley ended up in dire straits over the matter and was threatened with bankruptcy. Six months later and still unable to accrue the necessary funds to solve his problems, with his special diet appearing not to work and his unrelenting workload undiminished, Huxley was pushed to the verge of a nervous and physical breakdown.

Foreseeing Huxley's imminent collapse, Charles Lyell initiated a plan to help him and set about raising a fund to allow Huxley to take some time off work and to pay off his burdensome legal fees. Darwin was called upon to ensure that Huxley would accept the money and agree to taking an extended leave. Despite the fact that Emma thought Charles' method of writing to Huxley over the issue was more than a little clumsy, Huxley accepted the offer in the spirit it was made and immediately began to make plans for a lengthy foreign trip. In June 1873 he set off for a month-long walking holiday with Hooker and was then joined by his wife Henrietta and their children to spend the rest of the summer travelling.

While the Huxleys toured Europe, Charles, enjoying a rare period of glowing good health, travelled with Emma to visit her niece Effie (daughter of Hensleigh and Fanny Wedgwood), who had just married the politician Thomas Farrer. When they arrived they were greeted with a shock. Two weeks earlier on the sitting-room floor of the Farrers' new home a body had lain until it was borne back to London for an elaborate funeral. The dead man had been Samuel Wilberforce.

According to the still excited Farrers, Wilberforce had fallen from his horse in a nearby field and his companion, George Leveson-Gower, 2nd Earl Granville (leader of the House of Lords), and a servant had carried his body to this, the nearest house. During the two days in which Wilberforce had lain on the floor, the newly-weds had been visited by a succession of dignitaries, including the Prime Minister, Gladstone, who had travelled from London to pay his last respects.

Although Wilberforce had vocally protested against Darwin and his work he had borne him no personal animosity; for his part Darwin perceived Wilberforce as a rather misguided individual, whose undoubted intelligence was blunted by dogma. Not surprisingly, Huxley was less than magnanimous. When he heard eventually of the death of his old rival, he wrote to the

Irish physicist John Tyndall, declaring: 'For once, reality and his brain came into contact and the result was fatal.'[2]

Little more than a year after returning from his walking holiday in Europe with Huxley, Hooker was to face his own personal crisis. At the pinnacle of his career in November 1874, he was thrown completely off balance by the sudden death of his wife Frances, leaving him alone with six children, three of whom were still young. It was a devastating blow and pushed a middle-aged and stressed Hooker into an intense depression. He stayed at Down for a while and Darwin advised him to throw himself into his work as a remedy for his grief, offering him a new task to top off his already overwhelming administrative responsibilities – Darwin wanted his help in dealing with the latest attack from Mivart.

In the July 1874 issue of the *Quarterly Review*, Mivart had attacked viciously a piece written by Charles' son George and used the opportunity to have another dig at Darwin's own theories expressed in the *Descent*. In his piece, Mivart equated Charles and George Darwin with decadent and degenerate heathens. An understandably outraged Darwin read the piece and mobilised immediately his supporters against Mivart and the paper he was writing for. By the late autumn the argument had still not been settled. Mivart had published a half-hearted apology which only served to anger Darwin further and so he asked Hooker to intervene, knowing that he would relish the opportunity of counter-attacking the hated Mivart and at the same time it would help to distract him from his grief.

Rather than attacking Mivart openly, the X Club simply closed ranks. Darwin wrote Mivart a damning letter announcing in no uncertain terms that he would never speak or communicate in any way with him ever again and Darwin's supporters ignored Mivart and blocked any efforts he made to further his career. Such was the influence of the X Club that because of his ill-conceived and repeated attacks on Darwin, Mivart's career was effectively halted and he never acquired any of the establishment positions his talent would have allowed.

Better health for Darwin meant that during the early part of the 1870s he had the energy to socialise again. He had been a gregarious man when he was younger and in contrast to his later reputation as a rather severe reclusive figure, Darwin loved to be with people and to be entertained. It was always his health that prevented him from travelling and visiting friends and family rather than any difficulty with people or shyness. He enjoyed dinner parties and took delight in having house guests at Down as long as he was physically well.

In January 1874 he was even persuaded to attend a seance at his brother Erasmus' house in London. Although Darwin declared constantly his annoyance that seemingly intelligent people were being taken in by Spiritualism, the craze was still showing no signs of abating and a number of respected figures were becoming interested in the phenomenon. Wallace had long before been discredited in Darwin's eyes for his insistence on mixing science and mysticism, but, when the highly regarded statistician and eugenist Francis Galton attended a seance and came away expressing fascination if not actual conviction, Darwin succumbed finally and agreed to attend one himself.

It was a curious event. Erasmus, priding himself on his cosmopolitan reputation, had invited an eclectic mix of his associates. Galton was there, Emma's brother Hensleigh Wedgwood and his wife Fanny, Marian Evans (George Eliot), whom both Charles and Emma had long been anxious to meet (*Middlemarch* had been published in 1871–2), and in disguise, at Darwin's urgent request, the arch-sceptic himself, Huxley, newly refreshed by his travels and as sharp as ever.

The members of the party joined hands and fell silent, but nothing happened. After a while Charles began to feel sick; making his excuses he went upstairs to lie down while the others continued. After a while, from his room, Darwin could hear strange noises from below but stayed away, later hearing that the table had risen from the ground and that according to some guests it had almost touched the ceiling.

To Darwin's great relief Huxley was still not convinced and Galton continued to view the proceedings with scepticism. The problem was, at that point, neither Galton nor Huxley could prove conclusively that these mediums were frauds. The matter was only really settled to their satisfaction a few years later when, in October 1876, one of Huxley's students exposed a sensationalist travelling medium called Henry Slade. After a court case Slade was found guilty and sentenced to three months' hard labour. So pleased was Darwin that at least one trickster had been caught, he sent secretly £10 towards the prosecution's costs.

As Darwin grew older he moved consciously further and further away from theoretical work, almost as though with the delivery of his final general theory book, the *Expression*, he had said enough. Thereafter he devoted himself to experiment and books relating to these experiments. Even as early as the summer of 1872, before the publication of the *Expression*, he told Wallace: '[I] have given up all theories.'[3]

But although Darwin may have decided that he was now more interested in practical biology, conducting his experiments on plants and worms and writing about them, the world now saw him as the guru of revolutionary science. Some fifty years before Einstein became the most famous scientist in the world and purveyor of seemingly incomprehensible theories, Darwin, with his own apparently heretical ideas, was the embodiment of pure science and the most famous symbol of anti-establishment theorising. But the world was lagging behind. Darwin's days of radical scientific work were largely over. His intellect remained as sharp as ever during his final days; entering his late sixties, however, he wanted even less to be associated with any anti-establishment movement or philosophy, but it was impossible to escape completely.

What would today be called crank mail arrived almost daily at Down House. In letters from all over the world people asked him for his views on God and religion, requesting his endorsement of this or that theory, begging for his help in securing professional positions or to contribute to this and that scheme or project or experiment. And, as the public became increasingly fascinated with the man and his theories, the political establishment still considered it unwise to honour their country's most famous scientist. Gladstone paid a visit to Down in 1877 and spent an afternoon sipping Emma's tea and talking with Darwin in his study, but still no attempt was made to recognise publicly his contribution to the advancement of science. And if it had been, the proposal would undoubtedly have been blocked. Wilberforce may have died, but there were others more than ready to stop any attempts to honour the author of the *Origin*. Richard Owen would have been the first to try to destroy any such plans. A few years earlier he had tried to snatch Kew from Hooker's control and to place it under the auspices of his own British Museum, prompting Darwin to say of him: 'I used to be ashamed of hating him so much, but now I will carefully cherish my hatred and contempt to the last day of my life.'[4]

Honours did come from elsewhere. Although later attacked for not acting 20 years sooner, in 1877 Cambridge University finally bestowed upon Darwin their highest honour – an honorary doctorate. The awarding of the degree came as a great surprise to Darwin and he was naturally delighted. In relatively good health, he even managed to travel to Cambridge to accept the honour in person and to attend the parade through the city. Emma and half the Darwin family travelled with him and they all revelled in the pomp and ceremony surrounding the award. Huxley gave a speech at the Cambridge Philosophical Society in which he chastised the university authorities for not bestowing the honour decades earlier and the following day Darwin attended a special luncheon in his honour at Trinity College.

Meanwhile, elements within other academic institutions were hindering the awarding of special honours. In 1872, an attempt to elect Darwin as a Corresponding Member of the Zoology Section of the French Institute failed when the proposal received only fifteen out of 48 votes. This poor result was apparently caused by a small group of influential traditionalists who strongly disapproved of Darwin's work. One of them then even went as far as to write to *Le Monde* to protest against the idea of Darwin's election, stating:

What has closed the doors of the Academy to Mr Darwin is that the science of those of his books which have made his chief title to fame – the *Origin of Species*, and still more, the *Descent of Man*, is not science, but a mass of assertions and absolutely gratuitous hypotheses, often evidently fallacious. This kind of publication and these theories are a bad example, which a body that respects itself cannot encourage.

Six years later Darwin's supporters in France did manage finally to get their way by the convoluted method of putting Darwin's name forward in the Botanical Section of the Institute where he was voted a Corresponding Member by 26 out of a possible 39 votes.

Despite enjoying relatively good health, the Cambridge trip and the awarding of the honorary doctorate was a bright spell in an otherwise dark period in Darwin's domestic life. The 1870s had begun well; the older boys were beginning to forge careers and the younger ones were successfull academically, Etty was happily married and Elizabeth seemed contented at home with her parents. But then, troubles and tragedy started to become increasingly familiar.

Darwin's old mentor and lifelong friend Charles Lyell died at the beginning of 1875. He had been ill for some time and had gone into a rapid decline after the death of his wife Mary two years earlier. His death was really a relief from a collection of diseases which had blinded him and caused him immense pain. The relationship between Lyell and Darwin had cooled after Darwin had grown impatient with Lyell's inability to let go of the concept of the 'dignity of Mankind' but they had remained friends despite their differences. Darwin was asked to be a pall-bearer at Lyell's burial in Westminster Abbey, but declined not through any lack of love for the man but simply because he believed that he would probably collapse during the procession and send the coffin whirling off his shoulders. Lyell's death had been expected. He had led a full and comfortable life, he had contributed enormously to the development of geology and had been fulsomely recognised for it during his lifetime. For the Darwins there were far greater shocks just around the corner.

In spring 1876 Emma and Charles were delighted by the news that they were to become grandparents. When they visited their son Francis and his wife Amy, who had been married two years earlier, they were told Amy was five months pregnant. She went into labour on 7 September in one of the bedrooms at Down and a few hours later gave birth to a son they named Bernard. But then, soon after the delivery, Amy developed a fever, became seriously ill and died on the 11th.

It was the greatest tragedy to strike the Darwins since Annie's death 25 years earlier and it devastated the entire family. Charles had used his remedy of throwing himself into his work to overcome grief when coping with his own daughter's death, but in old age and with Emma and Francis in shock, this technique was barely enough to get him through the immediate crisis – he slid into a deep depression from which he never entirely emerged for the rest of his life.

In some respects it is difficult to understand why Darwin was in such a depressed state for most of his final years – so depressed that he even wrote to Hooker in 1875 expressing a semi-serious desire to commit suicide.[5] After all, he was free from all responsibilities other than those he placed upon himself for his own intellectual goals, he had no financial worries nor did he have now to worry about his children's future. In Emma he had a devoted and loving wife and they shared a beautiful home in an idyllic spot; he had achieved greatness, the respect of his peers and financial success. Certainly, he had not been honoured by the political establishment within his own country, but in some respects that was a mark of his achievement and was probably never a source of disappointment to him.

Ironically, the explanation for Darwin's depression may lie in the very fact that he was so comfortable and wanted for nothing. While Huxley and Hooker had to fight off political and scientific enemies, each to hold down several different jobs simultaneously, battle against financial problems and fulfil their responsibilities at the Royal Society, they probably had little time to get too depressed. Charles languished and fretted. All through his life he had been reliant on his intellect, it was the most important thing in his life. Then, as his physical health had improved, he began to worry that he would lose his intellectual powers. He tried to spot any decay in his working ability at every opportunity, as if, freed from physical pain, he had become a hypochondriac over his mental powers. When he wrote about this in 1881 at the age of 72, his anxiety is discernible:

> I am not conscious of any change in my mind during the last thirty years; nor indeed, could any change have been expected unless one of general deterioration. But my father lived to his eighty-third year with his mind as lively as ever

it was, and all his faculties undimmed; and I hope that I may die before my mind fails to a sensible extent.

The deaths of his daughter-in-law and his mentor Lyell were undoubtedly factors in his growing depression, but his life-style was also turning out to be his undoing. Having a comfortable home with more than enough material wealth was fine for a younger man with sufficient personal drive and self-motivation to overcome the temptation to lead a lazy, luxurious existence, but as he grew older he no longer had the energy for innovation. After writing the *Expression* there was really nothing more he could do with evolution. He had answered his critics using the limits of scientific knowledge of the day and had tried to find a fundamental mechanism for evolution with his theory of pangenesis. He had written three of the most important books on the subject of evolution and had pursued his goal to its limits. Now all he could really do was to delve into the minutiae of natural history and to tinker. This brought its own satisfaction, but his real *raison d'être* was fast disappearing.

Partial relief from this depressed state came from his supporters and disciples. Darwin was always gratified by the enthusiasm for his work shown by others and was boosted by their devotion to his theories. Haeckel continued to correspond and visited Down in October 1876, less than a month after Amy died. Emma was always thrown into a panic by the German's visits. Although she found Haeckel charming at times, there was no denying the fact that he was loud and excitable, the complete antithesis of the quiet, homely inhabitants of Down. Yet, in a peculiar way, the boisterous German helped Darwin during that autumn, relieving, for a short time at least, the mournful atmosphere in the house.

With the death of his young wife, Francis was at first confused and then cast into a lingering depression and it was decided that he should remain at Down with his parents. When he recovered from his loss and was able to work again, he became a close collaborator on many of his father's projects and even had his own study and laboratory built at Down – the first major extension to the property in 30 years. Francis had studied medicine at Cambridge, but true to the Darwin tradition initiated by both his father and his uncle, he displayed very little interest in pursuing it as a career and was drawn instead towards pure science, and botany in particular. Francis was not only keen but knowledgeable and was his father's closest assistant from the mid-1870s until Charles' death.

Second only to Francis Darwin was another acolyte, George Romanes. Romanes had been a friend of Francis at Cambridge and was a star student, winning a prize for his undergraduate essay *Christian Prayer and General Laws*. Genteel and highly intelligent, he was as enthusiastic about evolution as Haeckel and a perfect devotee to take evolution into the next century. The young man had the support of Huxley and recommendations from Hooker, for whom he worked at Kew, as well as praise from Francis, so that Darwin himself was soon won over. Suppressing the pain still felt over Mivart, Darwin invited Romanes to help him edit his later books and to work with Francis tending experiments at Down.

Romanes was a life-long supporter of evolution via natural selection and remained intimate with Darwin, helping him enormously during his final years. But, although he taught evolution to a generation of students at University College, London, Romanes was sadly never able to carry his mentor's work into the next century because he died in 1894 at the age of 46.

Perhaps motivated by an increased awareness of his own mortality exaggerated by hypochondria, in 1876 Darwin began a collection of autobiographical writings. Starting as an extended letter for Emma and the family to read after his death, these writings grew into a short collection of snapshots from different stages of his life. To historians and biographers, these recollections are rather disappointing. Throughout, Darwin was circumspect about his early life, giving mere flashes of events from his earliest recollections to the voyage aboard the *Beagle*; although they provided the basis for early biographies, they reveal little about the man's personality or feelings as a youth. This is all because, according to Darwin, he could remember little of his childhood; but the reader is left with the distinct feeling that he always considered discussion of his work to be far more important than descriptions of mere events in his childhood or efforts to portray his emotions at various stages of his life. He was most at home discussing his books, his views on religion and descriptions of his family and the people he had met during his lifetime – all perhaps a reflection of his own genuine modesty. This apparent oversight might also be interpreted as a need to avoid delving too deeply into memories of a sometimes painful childhood.

These autobiographical efforts absorbed him on and off for some years and were collected together after his death by Francis, who combined them with his own reminiscences and a collection of his father's correspondence to produce a book called *The Life and Letters of Charles Darwin*, first published in 1887. Much less successful were Charles' own attempts at biography.

Early in 1879, angered by the recently published *Evolution Old and New* written by a new enemy, one Samuel Butler, Darwin decided to put the record straight. Butler had claimed that evolution was not a creation of Charles Darwin and that Erasmus had said it all two generations earlier. Charles had never taken credit for the creation of the theory of evolution and had always acknowledged his grandfather's contribution. Coincident with this attack, Darwin came across an article about Erasmus in a German magazine called *Kosmos*. He contacted the author and suggested paying for the article to be translated and extended into book form in English and that he would write a biographical piece to accompany it. Excited by the thought of taking his revenge on Butler and inspired by the idea of trying something new, Darwin started the project with great enthusiasm but soon dried up. Biography simply was not his forte and Etty, as his literary adviser, told him so. After a great deal of reworking, which took him the best part of the summer, and bemoaning constantly the fact that he had started the project at all, he finally delivered what he and the family still considered a less than satisfactory biography of Erasmus to John Murray. Murray seemed to be the only one interested in the project, and agreed immediately to print 1,000 copies on spec and to split the profits, but, despite Darwin's international fame, the book was a flop, for once confirming the author's own pessimistic estimation of his efforts.

Indulging himself with sporadic forays into autobiographical writings, Darwin continued with his usual juggling act into the new decade, working on several things at once. Worms still fascinated him and he was progressing well with his account of their behaviour, and at the same time, with help from Francis and George Romanes, he had almost finished *The Power of Movement in Plants*, but then, two months into the 1880s, domestic troubles intruded again.

In February his second cousin and intimate friend since shared days at Cambridge, William Fox, died at the age of 75. Charles was too depressed and physically unwell to attend the funeral. Instead he buried himself further into the final chapter of *Movement in Plants* and blanked out the pain of loss. A month later another relative, this time Emma's brother Josiah Wedgwood, died aged 85. Again Charles was too low to attend the funeral and Emma also stayed at home, mourning her brother alone in her room in Down House.

Despite the depression which had been with him now for some five years and the deaths of several elderly members of the family, Charles was able to continue working through it all. He knew time was running out and although he had no more great theories or earth-shattering discoveries in him, he wanted to deliver the books he had started and to complete his experiments into both animal and plant behaviour. Able now to pass on some of the more mundane aspects of his work to Francis and to George Romanes, he could move faster. He managed to deliver the final draft of *Movement in Plants* in May 1880, and up until the last few months of his life, Darwin was still working at a pace at least equal to that of his middle age when physical illness hampered him far more.

At the same time he did not forget his friends. Towards the end of 1880 he discovered that his old colleague Wallace was going through hard times. Never the most responsible man with money matters and having entered into several unwise investments, by the middle of 1880 Wallace was about to be made bankrupt. A group of his friends rallied support for him and came forward with a proposal to acquire a state pension for Wallace and his family as a reward for his enormous contribution to the advancement of science. They approached Darwin with a request to put a word in for him, which he was more than happy to do. Darwin wrote to the influential Duke of Argyll and called upon Huxley, Hooker, Lubbock and others to help. Thanks in no small measure to Darwin, the plan worked and early in 1881 Wallace was awarded an annual pension of £200. Ironically, the award was made the same week that Emma completed her accounts for the year 1880 and the Darwins tried to decide how best to re-invest the £8,000 surplus they had earned from their many investments during the previous twelve months.

Between experiments and writing his final books, Darwin continued to visit his friends and colleagues in London. In fact, during 1880 and the early part of 1881, he socialised at least as much as he had done when he was struggling with the *Origin* and the *Descent* in his younger days. He visited Huxley and Hooker and even returned to Cambridge, but agreed to the visit only after Emma had arranged a private train to take them to Cambridge from the nearest station, Bromley, without the need to change trains in London.

Ostensibly to thank him for helping in the Wallace affair, in February 1881 Darwin called upon one of his most vehement critics, the Duke of Argyll. Meeting for the first time, the two men got along famously. The conversation was friendly and they even discussed politics and religion – subjects over which they eventually agreed to differ. Darwin was asked if he could say honestly, face to face, that he believed that the great beauty of Nature was really a chance thing. He responded by saying that he understood the difficulty in accepting such a thing, but yes, he did believe evolution worked without design.

The weather during the summer of 1881 was idyllic and the Darwins enjoyed many happy, sunny days at Down. According to later accounts from his children, for a time Charles had snapped out of his depression. In future years they reminisced about sun-drenched afternoons with their parents in the garden, enjoying picnics on the lawn and of their father lying on the drawing-room sofa, listening contentedly to Mozart being played on the piano. But this was to be the last period of good health and therefore of extended work and socialising in Darwin's life. In early August 1881 he dined with the Prince of Wales (later

Edward VII) and other dignitaries in London and the same week he sat for his portrait to be painted by Huxley's son-in-law John Collier, commissioned by the Linnean Society, but Darwin could sense that time was fast running out. He had just completed his final book and was tinkering with his autobiography, his energy oozing away when on 26 August, Emma answered the door at home, accepted a telegram and walked into Charles' study to read it to him. His brother Eras had died in London.

To Charles, the death of his brother was final confirmation that he too had reached the end of an era and it threw him into a physical and emotional nadir from which he never fully recovered. Despite the enormous strain, he supervised not only arrangements for his brother's funeral, which took place in Downe, but the sale of Eras' house and the details of his will; the effort took its toll.

The heart palpitations from which Darwin had suffered from time to time for many years became far worse and he was diagnosed as suffering from angina. With it came other physical symptoms – he became very tired early in the evenings and lost his appetite. He was finding it almost impossible to work and his final writings moved ahead at a snail's pace.

During a brief visit to London with Emma just before Christmas 1881, Charles decided to pay a visit to Romanes. Making his way slowly on foot and alone, he arrived at the house to find only the butler at home. The butler realised that the elderly man at the door was exhausted and asked him in, but Darwin declined and turned to go. Concerned, the butler watched Darwin totter along the road, stop and clutch the railings of a house and then faint on the pavement.

Romanes' butler immediately called for help and Darwin was taken to Etty's house where he and Emma were staying. He made a rapid recovery, but Emma was naturally shaken and could not get him back to Down House fast enough. From then on, she never let up her vigil and watched over him like a mother hen, pampering him and attending to his every need. For Emma, Charles' decline filled her with more than the usual dread of loss and the fear of being alone. Tortured still by the belief that her husband's dismissal of Christianity would mean that their parting would be permanent, she was terrified by the thought of his death. She had fretted and worried about this throughout their lives together. The talks she and Charles had had back at Maer, so long ago, had never really salved her fears. Although Charles had done as much as he could to protect her sensibilities, he had never been able to look her in the eye and say that he believed in the gospels, the Christian faith or the existence of an after-life. He remained convinced that death was the end whilst Emma wholeheartedly believed the opposite. The problem was, she was also convinced that if he did not believe, he would never make it to heaven and she would lose him for ever.

Darwin's decline came on suddenly, soon after his brother's death, but the end was slow. He was bedridden for long periods and then perked up enough to venture out and even to contemplate a little work. His worm book had been published in October 1881 and had sold remarkably well, a further pleasing confirmation that whatever the area of science he wrote about, his name would ensure success. It was his final complete work and Darwin was particularly proud of it. He still managed to answer letters and even to entertain guests at Down during the final months of his life. Members of the family visited as often as they could and he and Emma even managed to put up with a group of passionately atheistic writers who visited Down one afternoon in September 1881 to pay their respects to their bemused guru and to stay for tea. Apart from close family and friends, they were probably the last people to visit Darwin. He suffered another seizure in March 1882, this time alone along the Sandwalk. He was petrified that he would die there and then, away from Emma and the familiar surroundings of the home he had lived in for so many years. Somehow he managed to stumble back to the house where he collapsed in Emma's arms.

Even then family visits and conversation with Emma cheered him and brought back a sparkle to his eyes as he sat up in bed with a mass of pillows to support him. Charles was tenderly appreciative of all Emma had done for him. During his final weeks he told a family friend how he felt about his wife. 'What a miserable man I should be,' he said, 'without this dear woman.'[6] He knew he was slipping away and he was ready to die – the pain caused by the angina and the return of his immune-system problems was distressing to witness and agony to live through.

The end came on the afternoon of 19 April 1882. He had been bedridden for several days, the family had been summoned and they had taken turns watching him around the clock. Doctors came and went, prescribing mostly useless remedies to dull his pain. Then, after coughing up blood, and fainting for short periods only to regain consciousness and suffer more, he called for Emma who was resting in the next room. She rushed to his side and forced him to sip some whisky. It revived him for a moment but then his head dropped to Emma's breast; she cradled his head and he died.

William Darwin, the new head of the family, immediately took on responsibility for his father's funeral arrangements as well as the care of his distraught mother and siblings. The family had intended to have Charles buried in the village in which he had spent the last 40 years of his life and had begun to make preparations. But behind the scenes, and at first unbeknownst to the family, actions were being taken which would result in an altogether different burial.

Darwin's cousin and friend Francis Galton, in alliance with Huxley, was the first to initiate a plan to have the great scientist buried in Westminster Abbey. Galton and Huxley pulled strings at the Royal Society and through the auspices of the politically influential team of Thomas Farrer and John Lubbock, they managed to generate enough support from MPs to sign a petition to be sent to the Dean of Westminster. At the same time Farrer tackled the Dean personally and through the X Club's media contacts the papers made a big show of the fact that, if the country had been unable to honour one of its greatest figures during his lifetime, then it must not fail to do so after his death.

Meanwhile, the feelings of the Darwin family had not been forgotten. William Spottiswoode, a member of the X Club and close friend of the Darwins, approached Emma and William. Although initially stunned by the proposal, they gave their consent as long as it did not meet with opposition and it was made clear that the suggestion had not come from them.

The plan was successful, the Dean gave his approval and the date for Darwin's burial in Westminster Abbey was set for 26 April. Although neither Queen Victoria nor Mr Gladstone attended the funeral of one of Britain's greatest scientists, Charles Darwin was well represented by both his friends in political circles and most importantly the scientific establishment. Late during the morning of the 26th, the House of Commons emptied and the committee meetings of learned societies were adjourned and members made their way to Westminster Abbey to pay their last respects.

Only rarely has such an illustrious group of pall-bearers been seen – even in Westminster Abbey. The group of ten men included two knights, Sir Joseph Hooker and Sir John Lubbock, two dukes, including the Duke of Argyll and an Earl, as well as William Spottiswoode (then President of the Royal Society), one of Darwin's closest friends and Spottiswoode's successor, Thomas Huxley and the co-discoverer of the theory of evolution by natural selection, Alfred Russell Wallace.

Conspicuous by his absence from the honours list during his lifetime, Darwin was at least given a burial befitting his status and his enormous contribution to the advancement of human understanding. Even then the honour of burial in the most august church in England, beside the great statesmen and leaders as well as many less than successful members of the Royal Family who had contributed very little, happened only because of Darwin's devoted friends and colleagues. If Darwin had known that he was to be buried in Westminster Abbey, he would probably have considered it rather amusing, caring as little as he did for such man-made hierarchies and even less for the importance of what was to happen to his body after his mind and personality ceased to exist. What would have been far more important to him would have been the knowledge that his work had survived him, that he was absolutely right about evolution and that future scientific discoveries would further validate his work, that his theory would stand as one of the great achievements of intellectual endeavour and would be his lasting epitaph.

Notes

Unless otherwise specified, all quotations in this chapter are from *The Autobiography of Charles Darwin and Selected Letters*, 3 vols, ed. Francis Darwin, John Murray, London, 1887.

1 H. Atkins, *Down: The House of the Darwins: The Story of the House and the People who Lived There*, Royal College of Surgeons of England, 1976, p. 99.
2 Letter from T. H. Huxley to J. Tyndall, 30 July 1873, T. H. Huxley papers, Imperial College of Science and Technology, London.
3 A. R. Wallace, *My Life: A Record of Events and Opinions*, Chapman Hall, London, 1905, vol. 2, p. 90.
4 *Life and Letters of Sir Joseph Dalton Hooker*, ed. L. Huxley, John Murray, London, 1918, vol. 2, p. 131.
5 *The Life and Letters of Charles Darwin,* ed. Francis Darwin, John Murray, London, 1887, vol 3, p. 328.
6 R. Colp, Jnr, *To Be An Invalid: The Illness of Charles Darwin*, University of Chicago Press, 1977, pp. 93–4.

Chapter 14

Evolution after Darwin

As we have seen, at the time of Darwin's death, 'Darwinism' was in decline. It would be another half-century before natural selection became established as the key mechanism involved in evolution, a centrepiece of what became known as the 'modern synthesis', a combination of Darwin's ideas with the developing twentieth-century understanding of the mechanism of heredity. Two decades after that, in the 1950s, came the identification of the molecule which carries the genetic message, DNA; and even in the 1990s, more than a hundred years after Darwin's death, evolutionary studies are still providing new insights, especially into human origins and human behaviour. But many of the lines of investigation which were to prove crucial in the rehabilitation of Darwinism, decades after Darwin's death, had already begun while he was still working on the *Origin*.

One essential piece of the puzzle was the development of an understanding of the way cells themselves work. Living things are made of cells, and new individuals (whether plants or animals) result from the fusion of two special cells, one from each parent. From this single fused cell, the new individual grows to become a tree, or an elephant, or a barnacle, or a human being. One of the insurmountable problems of Darwin's pangenesis hypothesis was how to get the information about environmental influences on the body of an adult individual into the special cells involved in reproduction.

Growth of an individual from a single fertilised cell involves just the opposite problem – how information about the nature of the body that is being built up gets out of the original cell and into all the new cells that are being manufactured. But even in the 1850s it was already clear that this is a more natural process. In 1858, a year before the *Origin* was published, the German pathologist Rudolf Virchow (1821–1902), working at the University of Berlin, showed that every living cell is derived from a pre-existing living cell. Each cell consists of a bag of watery jelly, with a central concentration of material known as the nucleus. By the middle of the 1870s, while Darwin was still active, it was clear that in order to make a new cell, the material of the nucleus divides itself into two portions, which move to opposite sides of the cell. The cell itself then divides, forming two new cells each with their own nucleus.

It is this process of division, repeated many times, that has made each of us out of a single fertilised egg. During the process of fertilisation, and only at that time, the process of cell division seems to be reversed, as two cells merge and their nuclei combine.

This line of study was taken up in the 1880s by another German researcher, August Weismann (1834–1914), working at the University of Freiburg. He realised that the hereditary material (known at the time as 'germ plasm') is carried in the nuclei of cells. The sperm and egg from the two parents each contribute to the nuclear material of the fertilised cell from which a new individual develops, and somehow copies of this hereditary material are passed on into each new cell subsequently formed by division. In 1879 another German, Walther Flemming (1843–1915), had discovered that the nuclear material contains thread-like structures which readily absorb the coloured dyes used by microscopists to stain the parts of cells and make them visible; these threads became known as 'chromosomes', and Flemming showed that during the process of cell division it is chromosomes that are shared between the two 'daughter' cells to make new cell nuclei.

Weismann realised that the chromosomes must carry hereditary information, and that 'heredity is brought about by the transmission from one generation to another of a substance with a definite chemical and, above all, molecular constitution'.[1] He called this chemical 'chromatin', and explained that whereas in the kind of cell division associated with growth and development the chromosomes are duplicated before the cell divides, so that each daughter nucleus has a full complement of chromatin, in the special kind of cell division that produces egg and sperm cells the amount of chromatin is halved. Only when the two sex cells fuse is the full complement of chromatin restored and the potential for a new individual created.

One of the key discoveries that Weismann made came from microscopic studies of developing embryos (they were actually those of small marine creatures, but the conclusions are quite general). He showed that the cells which will eventually develop to become the organs responsible for producing egg or sperm cells by this special process are set aside from the rest of the developing body quite early on (indeed, all organs are 'specialised' in this way early during development, although the reproductive cells specialise very early on). The cells responsible for reproduction do not contribute to the growth of the rest of the body at all, but concern themselves only with reproduction. Equally, the rest of the body does not contribute to the development of the reproductive equipment. So there is no way that outside influences can affect the production of chromatin to be passed on to the next generation.*

By the time his final book appeared in 1904, Weismann had proved that pangenesis, Lamarckism, and the notion of everyday environmental influences directly causing variation were all wrong. By then he had become the most ardent Darwinist of them all (certainly more ardent than Darwin had been in the last 20 years of his life), arguing that natural selection was all that was required to explain evolutionary change. But he argued the case so adamantly that he roused the Lamarckians to spirited opposition, and at the end of the nineteenth century, although there were two schools of thought about evolution, in

---

* In extreme cases, for example intense radiation, it is possible to damage the DNA in the sex cells and thereby to affect the next generation, as the chilling aftermath of the Chernobyl nuclear accident made clear; but this is not the kind of outside influence Darwin or Lamarck were thinking of.

spite of the evidence from cell biology, the Darwinists were still in the minority. The next twist in the tale gave them, initially at least, another set of opponents to contend with; but in the end it would be the Lamarckians who would be routed by the new studies of heredity at work.

Again, we have to pick up the thread of the story by going back to the 1850s, this time to look at the work of Gregor Mendel, mentioned in Chapter 8. Mendel was born in 1822, in what was then Moravia, a region lying across the borders of modern Germany, Poland and the Czech Republic. He was baptised Johann, and took the name Gregor when he became a novice monk, in 1843. As the son of poor peasants living in a tiny village, he was lucky to receive any kind of an education, but was such a promising student that his younger sister Theresia renounced her share of the modest family estate in order to help to pay his way through college until he was 21. Theresia's selfless help did not go unrewarded, and the happy ending to this tale of devotion is that in due course Mendel provided similar help for her three sons.

In 1843, with the modest family resources exhausted, there was only one possible way in which the young Mendel could have any hope of continuing his studies even part-time, and he duly joined the Augustinian monastery at what was then Brünn (the capital of Moravia), now Brno in the Czech Republic.

Mendel soon became established in the religious community at Brünn, partly because several of the older monks died over the next few years so that he rose rapidly up through the ranks. He was ordained in 1847 and after a short spell as a parish priest became a schoolteacher at a country town in southern Moravia in 1849. He was a successful teacher in spite of his lack of formal training, and this encouraged the Prelate of the monastery to send the young priest to study at the University of Vienna, which he attended from 1851 to 1853. From 1854 to 1868 Mendel taught at a new technical school, the Brünn Modern School, before himself being elected Prelate. It was while he was teaching at the Modern School that Mendel carried out the studies for which he is now famous; but although he lived until 1884 (just two years after Darwin died), his scientific work ended when he became the head of the community of Augustinians in Brünn and, as has been the fate of so many good scientists, research had to give way to administrative duties.

Even this brief sketch of Mendel's life should be enough to dispel the illusion that he was some kind of hermit who lived completely cut off from the rest of the world. Vienna in the 1850s was hardly a backwater of civilisation (although the young priest spent only two years as a student there), and in 1862 he travelled with a party of tourists to the Great Exhibition in London; according to one historian, a photograph of Mendel taken at the time shows 'not so much the recluse as a jovial Friar Tuck'.[2] Alas, as far as historians have been able to ascertain, at the time of his visit to England Mendel did not know of Darwin's work, although among the books he left when he died was a marked copy of the German edition of the *Origin* published in the year of this visit to London. It is certain that the two pioneering biologists never met – Darwin's movements are known precisely from his own records, and they were never in the same town at the same time.

By then, three years after the publication of the first edition of the *Origin*, Mendel was six years into his own research, which concerned the way characteristics are inherited by pea plants. He was not the first naturalist to make investigations of this kind (and he certainly would not be the last), but he carried out meticulous studies over several generations, keeping careful records and analysing the results in detail. For this he is rightly regarded as the father of genetic research.

Mendel's experiments took seven years to complete and involved about ten thousand individual pea plants of six different varieties. He chose peas to study precisely because they do have distinctive characteristics which show up in different individuals in a simple way – for example, some plants are tall and others short, while some peas are wrinkly and others smooth. He looked exhaustively at the various possible combinations of patterns of variation, and the way in which the characteristics were passed on from one generation to the next, checking to see whether, for example, the offspring of a cross between a plant with wrinkly peas and a plant with smooth peas produces smooth or wrinkly peas. Crucially, he then went on to find out what kind of peas are produced in subsequent generations (the 'grandchildren' and later descendants of the original plants).

His key discovery was that a propensity for each kind of characteristic is passed on by each parent to the next generation, even though the expression of the characteristic may show up as only one of the two possibilities. The 'suppressed', or recessive, version of the characteristic can still show up in later generations, having been passed on unaltered even though it was not expressed in the intervening generation(s) of plants.

One example should make this clear. Mendel took plants from a variety which always produces green peas, and plants from a variety which always produces yellow seeds. When he cross-fertilised them, he did not get peas with an intermediate colour, as the old idea of blending inheritance would have suggested. In fact, in this example all of the peas in the next generation were yellow. The 'greenness' seemed to have disappeared. But when those yellow seeds were planted and grew into plants that were allowed to fertilise themselves naturally, they produced a generation of 'grandchildren' in which three quarters of the seeds were yellow and one quarter were green.

The actual numbers in Mendel's experiments were 8,023 individual peas, 6,022 of which were yellow and 2,001 green. Each individual pea pod might contain five or six yellow peas and two or three green ones. When this generation of peas were planted out in their turn, the green ones produced only green ones, but the yellow ones showed a more complicated pattern. Out of 519 yellow seeds, 166 produced pods in which all the peas were yellow, while the rest showed the pattern of the previous generation, yellow and green peas in the ratio 3:1.

In fact, Mendel's published results are so close to the 'ideal' ratio 3:1 that they are almost too good to be true. Statistically speaking, with only a few thousand peas to work with, Mendel shouldn't have got such good results. There is a suspicion that he may have tidied up his results to make them more nearly fit round numbers; whether or not that may be the case, experiments like these have been performed many times since, and do show the overall 3:1 ratio that he discovered.

The explanation of this pattern of inheritance is that the colour of the peas is determined by something passed on as a unique piece of information from parent to offspring. There must be a 'factor' for greenness (G) and a factor for yellowness (Y). Each individual carries two copies of the factor – in the light of Weismann's work, each *cell* carries two copies of the factor. If both these copies specify greenness (GG), the peas are green; if both copies of the factor specify yellowness (YY), the peas must be yellow. But if the plant's cells carry one copy of each factor (GY), the greenness is depressed entirely, and the peas are yellow. The factor for yellowness is said to be dominant over its counterpart for greenness.

Now you can see how heredity works. When the cells divide to make the reproductive cells, only one copy of each factor goes into each sex cell. In the original green peas, this meant that every sex cell carries the factor G, because all the parents were GG. But the first generation of yellow plants must all have been YY, so all their sex cells carried the factor Y. In the next generation, every pea plant produced by crossing the two original varieties, carried both factors, GY, but all the peas were yellow because the Y factor dominates. When the flowers on these pea plants fertilise each other, however, there are now four possibilities.

In the next generation, an individual plant might inherit the factor G from both parents, ending up with a GG combination and producing green peas. Or it might end up with a Y from each parent, ending up with a YY combination and producing yellow peas. But it could also inherit a Y from one parent and a G from the other, forming a YG combination and therefore producing yellow peas. And it should not be forgotten that there are two ways to achieve this last possibility, since the plant could inherit G from the first parent and Y from the second, making a GY combination which has the same effect as YG. Add up all the possibilities, and just one quarter of the grandchildren will be green, while three quarters will be yellow – exactly as Mendel found.

These studies showed that characteristics are passed on intact from one generation to the next. Instead of your being a blended combination of an equal one-eighth mixture of each of your great-grandparents, you inherit one eighth of your characteristics from each great-grandparent, but you can inherit them intact, so that it is possible for a child to have 'his grandfather's nose' or 'her mother's eyes'.

In most cases, the patterns of inheritance are much more complicated than the inheritance of colour in peas. Many different characteristics interact with one another in complex fashions so that it is impossible to say that there is 'a gene' (to use the modern terminology) for, say, height in human beings, or nose shape. In addition, the way characteristics are expressed depends on the environment in which the individual grows up – whether he or she has enough food, for example. But the essence of heredity is as Mendel discovered – factors which specify particular characteristics are inherited in pairs, one from each parent, but only one of which is actually expressed.

Mendel's results were presented to the Brünn Society for the Study of Natural Science in February 1865, and published a year later in its *Proceedings*. Copies of the *Proceedings* automatically went to academic libraries in other cities, including London, Paris, Vienna, Berlin, St Petersburg, Rome and Uppsala. In addition, Mendel sent copies of the paper (reprints) to about 40 individual scientists that he thought might be interested in the work. Nobody took much notice, though the paper was not completely ignored and did get a passing mention in an article in the *Encyclopaedia Britannica* and other more obscure publications. According to Mendel's biographer, the only real impression the work made at the time was on the Church authorities in Moravia, who suggested that evolutionary studies were not quite the right thing for a monk who wanted to get on in his profession. Their response may have encouraged Mendel to retreat into his monastic work.[3]

One reason why Mendel's work made little immediate impact is that it came just before the discovery of chromosomes and details of the way the cell divides. *After* chromosomes had been discovered, this kind of research could be more naturally slotted in to the development of ideas about heredity; and, of course, we now know that Mendel's 'factors' are carried by the chromosomes themselves. So it is no surprise that by the end of the nineteenth century, following Weismann's work, several researchers were carrying out similar studies to those of Mendel, and were about to rediscover the hereditary principles that he had investigated.

One of the key players in this phase of the development of the understanding of how evolution works was a Dutch botanist, Hugo de Vries, who had been born in 1848 and lived long enough (until 1935) to see the full flowering of Darwinism in its modern form. As early as 1889, just five years after Mendel died and seven years after Darwin died, de Vries published a book called *Intracellular Pangenesis*, in which he tried to adapt Darwin's ideas to the developing understanding of how cells worked. He suggested that the characteristics of a species could be separated out into a number of distinct units, each of them due to a single hereditary factor which was passed on from generation to generation more or less independently of the others. He called these hereditary factors 'pangens' (sometimes translated into English as 'pangenes'), from Darwin's term 'pangenesis', and in the 1890s he carried out a series of plant-breeding experiments, rather like those of Mendel, to see if distinctive characteristics associated with specific pangenes could be traced through the generations.

Similar studies were being carried out at about the same time by the British zoologist William Bateson, who lived from 1861 to 1926. He was particularly interested in the importance of variation to evolution, pointing out (quite rightly) that without variation there would be no raw material upon which natural selection could act. But Bateson went much further than Darwin in thinking that these variations could be quite large from one generation to the next – he was indeed, a proponent of the idea of saltations, although he looked for a natural explanation of these sudden jumps, rather than the hand of God at work.

Bateson and de Vries both came to the conclusion that there were two kinds of evolution at work, gradual changes which fitted species ever more tightly to their ecological niches, and sudden dramatic jumps that created new species. Some people still share this misconception, so perhaps we should say clearly here and now that they were wrong. True, in the normal course of events routine evolution generation by generation and year by year *does* tend to fit species more tightly to their niches rather than create new species. There are indeed more dramatic changes that lead to the creation of new species; but these are not changes in the way evolution works. Rather, they are caused by changes in the environment.

If individual members of a species get shifted, for whatever reason, to a new environment (as happened to the ancestors of the Galápagos finches), their descendants will be subjected to new evolutionary pressures and will become adapted to new ecological niches, becoming new species in the process. Or the individuals might stay in the same place, only to have the environment change around them – a new Ice Age may develop, shifting the evolutionary balance; or there may be a great disaster, such as an impact from outer space of the kind that killed off the dinosaurs. The fossil record shows that there was an explosion of evolution of new mammal species after the dinosaurs died out, some 65 million years ago; but this was not because natural selection was working any differently from before. It was because the dinosaurs were no longer there to compete with the mammals. The *potential* for rapid evolution always exists, but it is realised only when individuals are placed in different environments.

In 1899 de Vries was preparing his own work for publication, and checked back through the scientific literature in order to place it in context. It must have been with distinctly mixed feelings that at this point, for the first time, he discovered Mendel's work. On the one hand, here was independent confirmation of the main features he had discovered about the inheritance of distinctive characteristics; on the other hand, he had been pre-empted by an obscure Moravian monk working more than thirty years earlier.

De Vries published two papers describing his findings early in 1900. One, in French, was very short, and made no mention of Mendel. But the other, in German, was much longer and more thorough, and gives full credit to the Moravian monk, commenting that 'this important monograph is so rarely quoted that I myself did not become acquainted with it until I had concluded most of my experiments, and had independently deduced the above propositions'.[4] At the very end of the paper, he summed up:

From these and numerous other experiments I drew the conclusion that the law of segregation of hybrids as discovered by Mendel for peas finds very general application in the plant kingdom and that it has a basic significance for the study of the units of which the species character is composed.[5]

But if de Vries had mixed feelings about discovering Mendel's important monograph, imagine the frustration of the German botanist Carl Correns, who had also been carrying out plant-breeding experiments (some of them on pea plants), had also discovered the Mendelian laws of inheritance, and had also only then come across Mendel's paper. And then, before he could publish his results, he received a copy of de Vries' French paper. As if this weren't enough, an Austrian, Erich Tschermak von Seysenegg, discovered Mendel's work in much the same way at about the same time. Just as the scientific world had clearly not been ready for Mendel's discoveries in the 1860s, so, equally clearly, in 1900 the Mendelian laws of heredity were a discovery (or rediscovery) waiting to happen.

The fact that so many people rediscovered these ideas at about the same time may, indeed, have given Darwinism a greater boost than if only one person had made the breakthrough at the end of the nineteenth century. With de Vries, Correns and others all potential rivals for the credit of making the discovery, there could have been a bitter wrangle about scientific priority. But with everyone agreeing that the real credit belonged to Mendel, the potential dispute was defused in a gentlemanly manner.

But this only paved the way for other disputes. De Vries, Bateson and others saw the Mendelian process of inheritance as evidence that new species were not created gradually by an accumulation of small changes over a long period of time, but all at once, in a series of a few jumps. De Vries introduced the term 'mutation' to describe this process, and published a book, *The Mutation Theory*, at the beginning of the twentieth century (Bateson, incidentally, introduced the term 'genetics' for the study of heredity).

Now, there were three rival theories of evolution – Darwinian natural selection, involving small changes accumulating over long times, but originating through chance variations; neo-Lamarckism, involving small changes which accumulate more rapidly through the striving of individuals towards a goal; and the mutation theory involving sudden dramatic changes from one generation to the next, the creation of what were later called 'hopeful monsters' drastically different from their parents, which might or might not be better fitted for survival.

The resolution of the dispute was a long, slow process. One of the key developments was a series of breeding experiments with fruit flies (*Drosophila*), carried out by the American Thomas Morgan from 1909 onwards. The spirit of these studies was very much like that of Mendel, studying the way characteristics (in this case, such features as eye colour, or wing shape) are passed on from generation to generation. The great advantage of fruit flies is that they breed much more rapidly than peas, and so many generations can be monitored in a relatively short time – each fly is only an eighth of an inch long, and they produce a new generation in just two weeks.

About the time that Morgan began these experiments, the Dane Wilhelm Johannsen coined the word 'gene', shortened from the pangene of de Vries and ultimately derived from Darwin's word pangenesis, to denote the unit of heredity, and Morgan took up this term.

Although he started out very much a mutationist of the de Vries kind, Morgan found in a series of painstaking experiments that the variations among individuals are much more subtle than de Vries and Bateson proposed, and convinced himself that a combination of the Darwinian idea of natural selection and a much less dramatic form of mutations was the right picture. In a powerful example of the way science ought to be done, he started out with one point of view (opposed to Darwinian evolution), but convinced himself by experiments that he had been wrong. As a sceptic who convinced himself, through his own experiments, that Darwin was right after all, Morgan underwent a 'conversion' that carried a great deal of weight.

From about 1910 onwards a fusion between Darwinism and Mendelism began to occur, as Morgan's experiments showed further that the genes themselves are individual units strung out along chromosomes like beads on a string. But Lamarckism still had a certain popular appeal.

Even today there are still people who find it hard to accept that the story of evolution is one of 'blind chance'. They want to believe that there is some purpose to evolution, that the old image of a ladder of creation still means something. Lamarckism offers the idea that life can strive upwards towards more perfect forms by its own efforts, making something noble out of the otherwise brutal reality of the ultimately meaningless struggle for existence.

The extent of the popular appeal of Lamarckism in the first quarter of the twentieth century can be seen in two works by George Bernard Shaw, *Man and Superman* (from 1903) and *Back to Methuselah* (from 1921). Shaw promotes the idea of 'creative evolution' rather than a series of senseless accidents selected in a blind struggle for existence. Just at the time when scientists were at last coming to a proper understanding of the effectiveness of Darwinian evolution by natural selection, in the world at large Lamarckism was being promoted as a much more comfortable view than the struggle for survival.

In the 1920s, however, the final nail was hammered into the coffin of Lamarckism when much more sophisticated mathematical analyses of the way mutations can spread through a population were carried out by scientists such as R. A. Fisher and J. B. S. Haldane in England, Sewall Wright, in the United States, and S. S. Chetverikov, in the Soviet Union. Different varieties of the same gene are called 'alleles' (the term actually predates 'gene', and was coined by Bateson). If, being simplistic, we say that there is a single gene (a particular piece of a particular chromosome) that decides what colour a pea is, then the different versions of that gene are alleles. There may be many (possibly dozens) of different equivalent alleles for each gene present among the individual members of a species, all different versions of the same gene; but we shall pretend that in this case there are just two variations on the theme. There is an allele for yellowness, and an allele for greenness. Each pea plant carries two alleles, one inherited from each parent, which may or may not be the same. Fisher calculated, in his classic book *The Genetical Theory of Natural Selection*, published in 1930, that if a new allele, produced by a mutation from an old one, gives those animals or plants that possess it just a one per cent advantage over those that do not (imagine, for example, that blue peas turn out to be that much more efficient at converting sunlight into food, and one plant develops an allele for blueness by mutation from the allele for greenness), then the new allele will spread through the entire population within a hundred generations.

Evolution, according to this new picture (variously called 'the new synthesis' or 'the modern synthesis'*), *is* about mutations; but the mutations need only be small to do their work. Recently the power of evolution at work has been monitored directly by more experiments on fruit flies, carried out by Francisco Ayala, of the University of California at Davis. A large population of these flies, all descended from a single pair, was divided into two. Half were kept in a room at 16° c, the other half in a room at 27° c. They were otherwise treated exactly the same, and allowed to breed, at a rate of about ten generations per year. After twelve years, the average size of the flies kept in the cooler room was ten per cent greater than the average size of the flies kept in the warmer room. The two populations had diverged at a rate of 0.08 per cent per generation. Darwinian evolution can be *seen* going on before our very eyes; let nobody try to tell you that it is 'just an hypothesis'.

One of the most dramatic changes in our own evolution occurred when *Homo erectus*, with a brain size of 900 cc, became Neanderthal man, with a brain size of 1,400 cc. In the geological record, this seems to have happened in the blink of an eye, from one stratum to the next. But as Ayala points out, if we allow the change to have taken place at the same rate as the fruit flies changed, 0.08 per cent per generation, and if we assume a gap of 25 years between generations (probably ten years more than the actual gap for our ancestors), then the whole transformation

---

* The phrase 'new evolutionary synthesis' was introduced by Julian Huxley, grandson of Thomas Henry Huxley, in his book *Evolution: The Modern Synthesis*, published by Allen & Unwin in 1942. If we had to draw the line somewhere, the publication of this book, 60 years after the death of Darwin, could be said to mark the moment when Darwinism finally became established as the best explanation of how evolution works.

could have taken place over just 540 generations, or 13,500 years. Such changes are indeed too rapid to show up in the geological record, but they are still slow by human standards.

There is another way of looking at this, an example we particularly like. It was provided by the American biologist Ledyard Stebbins, in his book *Darwin to DNA*.[6] Suppose there was a climate change which made larger animals more successful (exactly equivalent to the way flies kept in the cooler room get bigger with the passing generations). How long would it take to turn a mouse into a supermouse the size of an elephant?

If we allowed 12,000 generations for the process, then the difference from each generation to the next would be too small to detect. But if we assume that each generation takes five years to reach maturity (somewhere between the actual generation gap for mice and the one for elephants), the whole process would take just 60,000 years. If this process really did happen, then a million years from now fossil hunters scouring the record in the rocks might find the remains of mice from the period before the climate change, and fossil supermice from the period after the climate change. But the chances of their just happening to find mice in the intermediate stages, corresponding to the interval of 60,000 years when the change was going on, is negligible. The emergence of any 'new' life-form in the space of less than 100,000 years is so quick that as far as the coarse-toothed comb of the geological record is concerned it is instantaneous. Yet from a human point of view, mice may indeed be evolving into elephants at this very moment, but we could not tell because the process is so slow.

Among other things, this puts a different angle on what we mean by Uniformitarianism. On the time-scale of geology, it is quite normal for major catastrophes such as Ice Ages and impacts of meteorites from space to occur, perhaps every few hundred thousand years or every few million years. What seem like rare and unusual disasters on a human time-scale are part of the uniform processes which operate on longer time-scales to shape our planet and to modify life on Earth. Even disasters like the death of the dinosaurs can be seen as in some sense routine events, fitting the overall pattern of Darwinian evolution.

One of Darwin's sons, Leonard, lived long enough to see his father's work triumphantly established at the heart of the new synthesis, dying in 1943 at the age of 93. By then, Mendelian inheritance, the importance of *small* mutations, and the central role of natural selection in evolution were all firmly established. Genes had been identified as pieces of chromosomes, and the way genetic information was passed on from generation to generation was becoming clear. But nobody yet knew exactly what genes and chromosomes were made of, and how the genetic information was stored in them. Just ten years after Leonard Darwin died, even that problem crumbled in the face of scientific investigation.

The story of how Francis Crick and James Watson unravelled the structure of DNA, the life molecule itself, has been told many times, and we do not intend to repeat it here.[7] Suffice it to say that it was discovered that chromosomes are made of DNA, the famous double helix, and that the information which tells a cell how to go about its work and maintain the processes of life is strung out along the molecules of DNA in a series of chemical units which in effect form a four-letter genetic code, or alphabet. This is very much in line with Weismann's ideas from the 1890s; DNA *is* Weismann's 'chromatin'. Mutations occur when pieces of this genetic code get scrambled when the genes are being copied, either by one letter replacing another, or by a chunk of DNA being accidentally cut out and spliced in somewhere else, or by a piece of chromosome getting inverted, and so on.

One of the beauties of all this is that since individuals carry two alleles for virtually all their genes, if one becomes scrambled into nonsense it does not matter, provided the other allele does its job.* The individual that carries the scrambled allele will still survive, even if it is no better off than any other member of the species. But on the rarer occasions that a mutation does produce an improved version of the gene, it can take over and give the individual it belongs to an immediate advantage.

The whole process is now so well understood that geneticists

---

* Of course, the only time any of this matters is when genes are being copied into the sex cells (sperm or egg) where they will be passed on to the next generation. A mutation that occurs in a cell on the end of your nose has no effect on the rest of your body, or on your children – pangenesis is *not* the way evolution works.

can manipulate DNA in their laboratories, changing the code and rearranging it in the process known as genetic engineering. This, in many ways, is the ultimate proof of how evolution works – if we understand the natural process well enough to be able to copy it ourselves, there can be no doubt that we really do understand it.

But this is not quite the end of our tale. To bring the story of evolution right up to date, it seems appropriate to look at how Darwin's ideas are now casting new light on human behaviour, as well as on our own origins.

One of the most important results from DNA studies carried out over the past two decades has been the discovery that the genetic material of human beings (the DNA) differs by only one or two per cent from the genetic material of other African apes, the gorillas and the chimpanzees. Among other things, this suggests that the common ancestor that we share with African apes walked the Earth as recently as four or five million years ago. It means that in evolutionary terms we are 99 per cent ape, and only one per cent distinctively human.[8]

All of this would surely have delighted Darwin himself, who appreciated, a hundred years before this kind of analysis could be carried out, that the structure of the human body itself makes clear our close relationship to the African apes.

The fact that the relationship is *so* close, however, lends further weight to Darwin's own belief that the right way to understand what makes people tick is to look at humans 'as a Naturalist would at any other Mammiferous animal'. In its modern incarnation, this kind of study is known as sociobiology, and it has come in for the same kind of abuse today that Darwin himself received for daring to suggest that human beings represent just one species among many, subject to the same evolutionary laws as all other forms of life on Earth.

Sociobiology can provide insights into the relationship between the sexes, the conflict between the generations, why people fight wars, and how advertisers lure us into buying their products.[9] In the 1990s the last vestige of any specialness about being human seemed to be our ability to speak, to communicate complex ideas (like the theory of evolution) so effectively to one another. But even this last vestige of uniqueness has crumbled in the face of investigations by researchers such as Steven Pinker, of the Massachusetts Institute of Technology, who describes in his clear and entertaining book *The Language Instinct*[10] how language itself has evolved by the Darwinian process of natural selection.

There is nothing left which can be thought of as making human beings special; but we therefore have the potential to understand ourselves better than human beings have ever understood themselves, and we have that potential in large measure thanks to Charles Darwin. We are indeed one Mammiferous animal among many, and we live by Darwin's rules.

Notes

1 Quoted by David Young, *The Discovery of Evolution*, Natural History Museum, London, and Cambridge University Press, 1992.

2, 5 Ronald Clark, *The Survival of Charles Darwin*, Weidenfeld & Nicolson, London, 1984.

3, 4 Hugo Iltis, *Life of Mendel*, Allen & Unwin, London, 1932.

6 Published by W. H. Freeman, San Francisco, 1982.

7 See John Gribbin, *In Search of the Double Helix*, Corgi, London, 1985.

8, 9 See Mary and John Gribbin, *Being Human*, Dent, London, 1993.

10 Published by Allen Lane, London, 1994.

Appendix 1

Darwin on Darwin

At the end of his autobiography, writing on 3 August 1876, Charles Darwin attempted to sum up the distinctive characteristics of his own personality that had enabled him to make his contributions to science. We can think of no better way to give a closing insight into the mind of Darwin than to use his own words:*

I have no great quickness of apprehension or wit which is so remarkable in some clever men, for instance Huxley. I am therefore a poor critic: a paper or book, when first read, generally excites my admiration, and it is only after considerable reflection that I perceive the weak points. My power to follow a long and purely abstract train of thought is very limited; I should, moreover, never have succeeded with metaphysics or mathematics. My memory is extensive, yet hazy: it suffices to make me cautious by vaguely telling me that I have observed or read something opposed to the conclusion which I am drawing, or on the other hand in favour of it; and after a time I can generally recollect where to search for my authority. So poor in one sense is my memory, that I have never been able to

* From Francis Darwin, editor, *The Life and Letters of Charles Darwin* (John Murray, London, 1887).

remember for more than a few days a single date or a line of poetry.

Some of my critics have said, 'Oh, he is a good observer, but has no power of reasoning.' I do not think that this can be true, for the *Origin of Species* is one long argument from the beginning to the end, and it has convinced not a few able men. No one could have written it without having some power of reasoning. I have a fair share of invention and of common sense of judgement, such as every fairly successful lawyer or doctor must have, but not, I believe, in any higher degree.

On the favourable side of the balance, I think that I am superior to the common run of men in noticing things which easily escape attention, and in observing them carefully. My industry has been nearly as great as it could have been in the observation and collection of facts. What is far more important, my love of natural science has been steady and ardent. This pure love has, however, been much aided by the ambition to be esteemed by my fellow naturalists. From my early youth I have had the strongest desire to understand or explain whatever I observed, – that is, to group all facts under some general laws. These causes combined have given me the patience to reflect or ponder for any number of years over any unexplained problem. As far as I can judge, I am not apt to follow blindly the lead of other men. I have steadily endeavoured to keep my mind free, so as to give up any hypothesis, however much beloved (and I cannot resist forming one on every subject), as soon as facts are shown to be opposed to it. Indeed I have had no choice but to act in this manner, for with the exception of the Coral Reefs, I cannot remember a single first-formed hypothesis which had not after a time to be given up or greatly modified. This has naturally led me to distrust greatly deductive reasoning in the mixed sciences. On the other hand, I am not very sceptical, – a frame of mind which I believe to be injurious to the progress of science.

My habits are methodical, and this has been of not a little use for my particular line of work. Lastly, I have had ample leisure from not having to earn my own bread. Even ill-health, though it has annihilated several years of my life, has saved me from the distractions of society and amusement.

Therefore, my success as a man of science, whatever this may have amounted to, has been determined, as far as I can judge, by complex and diversified mental qualities and conditions. Of these the most important have been – the love of science – unbounded patience in long reflecting over any subject – industry in observing and collecting facts – and a fair share of invention as well as of common-sense. With such moderate abilities as I possess, it is truly surprising that thus I should have influenced to a considerable extent the beliefs of scientific men on some important points.

Appendix 2

The Major Characters in Darwin's Life

The major characters in Darwin's life and what happened to them after his death.

Elizabeth Darwin (Bessy): Charles' youngest surviving daughter. After her father's death, Elizabeth lived with her mother. She never married and died in 1926.

Emma Darwin: Charles' wife. Emma was in mourning for several months after Charles died. With the help of her large family she gradually recovered and by the end of 1882 had bought a house in Cambridge called The Grove. Down House was retained, but Emma, Elizabeth and Francis (with his son Bernard) all moved to The Grove just outside the city. Emma lived to the age of 88, dying peacefully at The Grove in 1896.

Francis Darwin: Charles' son Francis married a lecturer from Newnham College, Cambridge, Ellen Crofts, in the summer of 1883; they had one daughter. Francis completed editing his father's *Life and Letters*, which was first published by John Murray in 1887. He became Reader in Botany at Cambridge University and died in 1925.

George Darwin: Charles' second eldest son. He and his wife, Maud du Puy, had two sons and two daughters. He became

Plumian Professor of Astronomy and Experimental Philosophy at Cambridge University and died in 1912.

Henrietta Darwin (Etty): Charles' daughter. Despite suffering many childhood ailments, Etty lived to the age of 86. She was the editor of the family letters and produced a book *Emma Darwin: A Century of Family Letters*, first published in 1904.

Horace Darwin: The Darwins' last child to survive into adulthood, Horace became a successful businessman and was Mayor of Cambridge in 1896–7. He and his wife Ida had three children, one of whom, Nora (Barlow), edited her grandfather's autobiography. Horace died in 1928, and Nora Barlow in 1989.

Leonard Darwin: Charles' son. Leonard retired as an army major in 1890, became a Liberal Unionist MP in 1892 and was President of the Royal Geographical Society in 1908–11. After his first wife Elizabeth died in 1898, he remarried in 1900, but had no children from either marriage. He died in 1943.

William Darwin: Charles' eldest son remained in banking until retirement and died in 1914.

Thomas Farrer: Had married Emma's niece in 1873 and was instrumental in securing Westminster Abbey as Darwin's last resting place. The Farrers' home Abinger Hall in Surrey was a favourite house for Charles to visit in his old age. Farrer died at the age of 51 in 1884.

Francis Galton: Friend and cousin of Darwin's, Galton laid the foundations of eugenics and wrote a number of books including *Hereditary Genius*. He died in 1911.

Ernst Heinrich Haeckel: Darwin's friend and most ardent supporter of Darwinism. He remained Professor of Zoology at the University of Jena proselytising Darwin's ideas. He died in 1919.

Joseph Hooker: One of Darwin's closest friends. Hooker remained as Director of the Royal Botanical Gardens, Kew, until 1885. He produced a seven-volume work *Flora of British India* (1872–97) and received the Order of Merit in 1907. Hooker died in 1911 aged 94.

Thomas Huxley: Gained the name Darwin's Bulldog because of his vocal defence of Darwinism. He was also (along with Hooker) Darwin's closest friend. A year after Darwin's death Huxley became President of the Royal Society and held the position until 1885. After serving this term of office Huxley did little scientific research and spent the rest of his working life involved with civic responsibilities and helping to establish guidelines for public education. He died in 1895. His grandsons Julian and Andrew became eminent biologists (Andrew winning a Nobel Prize) while their brother Aldous gained fame as a writer.

John Lubbock: Neighbour and friend of Darwin. Lubbock was also instrumental in organising Darwin's burial. He was created 1st Baron Avebury in 1900 and died in 1913.

St George Mivart: Critic of Darwin. Remained lecturer in biology at St Mary's Roman Catholic College and died in 1900.

John Murray: Darwin's publisher. Although John Murray died in 1892, the company John Murray Ltd, which had been founded by his grandfather John McMurray in 1768, survives to this day and is still run by the Murray family.

Richard Owen: One-time friend and later great rival and critic of Darwin. Owen was superintendent of the British Museum until 1884 and was knighted in the same year. He died in 1892.

George Romanes: Assistant and friend to both Francis and Charles Darwin. He remained a researcher and lecturer after Darwin's death and wrote two highly regarded books, *Animal Intelligence* (1882) and *Mental Evolution in Animals* (1883). He died at the age of 46 in 1894.

Alfred Russel Wallace: Co-discoverer of the theory of evolution via natural selection. Although he continued to explore alternative theories and mysticism for the rest of his life, he also remained a keen experimenter and writer on more orthodox subjects. He wrote a book called *Darwinism* in 1889 and was finally made a Fellow of the Royal Society in 1892. He received the Order of Merit in 1908 and died in 1913 aged 90.

Hensleigh Wedgwood: Darwin's brother-in-law and friend had little to do with the Wedgwood business but remained close to his sister Emma. He died in 1891.

If you would like to know more about the personal lives of Darwin's children as adults, and would like a glimpse of Emma Darwin in old age, we recommend the delightful memoir *Period Piece* (Faber and Faber, London 1952), written by Darwin's grand-daughter Gwen Raverat (the eldest child of George Darwin), who grew up in Cambridge in the last decades of the nineteenth century and the first decades of the twentieth, surrounded by an extended family of Darwins.

Further Reading

Books marked with an asterisk include more technical material than the present book. The rest are accessible to the general reader.

Mainly about Darwin

Nora Barlow (ed.), *The Autobiography of Charles Darwin* (Collins, London, 1958).

John Bowlby, *Charles Darwin: A New Life* (Hutchinson, London, and Norton, New York, 1990).

Peter Brent, *Charles Darwin* (Heinemann, London, 1981).

Francis Darwin, (ed.), *The Life and Letters of Charles Darwin* (John Murray, London, 1887).

(An abbreviated form of this title was published in the United States in 1892 by D. Appleton & Co and given the title *Charles Darwin, His Life Told in an Autobiographical Chapter and in a Selected Series of his Published Letters*. This was then published as *The Autobiography of Charles Darwin and Selected Letters*; Dover, 1958, which is still in print today).

Adrian Desmond and James Moore, *Darwin* (Michael Joseph, London, 1991).

Alan Moorehead, *Darwin and the Beagle* (Hamish Hamilton, London, 1969).

Christopher Ralling (ed.), *The Voyage of Charles Darwin* (Ariel/ BBC, London, 1982).

Mainly about Darwin's work and/or its implications

Charles Darwin, *The Origin of Species* (Penguin, Harmondsworth, 1968 and many reprints; this is a reprint of the classic first edition, 1859, of the *Origin*).

Charles Darwin, *The Origin of Species* and *The Descent of Man* (single volume reprint of *both* great works, published by Random House, New York, in their Modern Library series; no date given, but note that this is the sixth edition, 1872, of the *Origin*).

Linda Gamlin, *Evolution* (Dorling Kindersley, London, 1993).

Stephen Jay Gould, *Ever Since Darwin* (Burnett Books/André Deutsch, London, 1978).

John Gribbin, *In Search of the Double Helix* (Black Swan, London and Bantam, New York, 1985).

Mary and John Gribbin, *Being Human* (Dent, London, 1993).

H. L. McKinney, *Wallace and Natural Selection* (Yale University Press, 1972).

*D. R. Oldroyd, *Darwinian Impacts* (Open University Press, Milton Keynes, 1980).

*John Maynard Smith, *The Theory of Evolution* (Pelican, London, third edition, 1975).

David Young, *The Discovery of Evolution* (Cambridge University Press, 1992).

About Darwin's life and work

Peter Brent, *Charles Darwin* (Heinemann, London, 1981).

Jonathan Miller and Borin Van Loon, *Darwin for Beginners* (Icon Books, Cambridge, 1992).

Duncan Porter and Peter Graham, *The Portable Darwin* (Penguin, London, 1993).

Gerhard Wichler, *Charles Darwin* (Pergamon, Oxford, 1961).

Also of interest

Antony Flew, *Malthus* (Pelican, London, 1970).

Hugo Iltis, *Life of Mendel* (Allen & Unwin, London, 1932, reprinted 1966).

Jonathan Weiner, *The Beak of the Finch* (Jonathan Cape, London, 1994).

Jonathan Weiner's book is such an important contribution to the understanding of Darwinism that we include here a review of that book which one of us (JG) wrote for the *Sunday Times*:

There are still people around who think that evolution is

'just a theory', in the same sense that your uncle Arthur might have some crackpot idea about how to grow better roses. 'Where's the proof?', they ask. Proof there is, in plenty, but until now it has largely been buried away in technical journals and books, not readily accessible to doubting Thomases. But now here is a book that, as its subtitle suggests, describes evolution going on, in exactly the way that Darwin surmised, literally before the eyes of biologists. Deliciously, the species that can be seen evolving as they adapt to changing environmental conditions are the finches of the Galápagos islands, first made famous by Darwin himself.

The story of the twenty-year research programme that revealed evolution at work is told in a highly American, almost novelistic style by Jonathan Weiner. Just occasionally, he lurches too far towards creating a new genre, the science 'n' shopping format, as we are told, for example, that one of the scientists involved in the project, Rosemary Grant, 'wears an Icelandic sweater, a long blue Laura Ashley dress, and sandals. Sunlight slants in the bay window behind her, through the spreading green arms of a katsura tree with shaggy bark.' But mostly he manages to control himself, producing a book that reads as easily as a goodish novel while slipping information across adroitly.

The information is so spectacular that it hardly needs dressing up with long blue dresses and sandals. Rosemary Grant, her husband Peter, and their colleagues have returned season after season to the Galápagos islands since 1970, and literally know every one of the finches on one of the islands by sight. They have kept family trees of the birds going back all that time, and know which birds have bred successfully and which have not. And they have trapped all but one or two of the birds, measuring and photographing them before releasing them back into the wild. They have watched populations decline in times of drought and boom in times of plenty; and they have seen how a change in the length of beak of a bird of less than a millimetre can make all the difference between it thriving and leaving many descendants or dying before it has a chance to reproduce. Only the biggest birds with the biggest beaks, able to cope with the toughest seeds, made it through the worst droughts.

Mixed with the page-turning descriptions of this field work under the harsh conditions on the island there is a wealth of information about Darwin himself and his theory of evolution, and the story is brought bang up to date, in technological terms, with the work of Peter Boag, who has studied the DNA in samples of blood from the finches, and can actually see the differences in the genetic code which specify the different designs of beak and make one finch good at eating seeds while another specialises in sipping nectar from cactus flowers.

All of this would be well worth the price of admission on its own, but Weiner has yet more drama up his sleeve. In the final chapters he discusses the continuing resistance to Darwinian ideas, and the extent to which even among scientists non-biologists sometimes fail to comprehend how evolution works. Chemists who invented new pesticides were astonished when populations of insects developed resistance to them, but as Weiner spells out the whole point about evolution is that any new method of killing things will, unless it completely wipes out a species, give rise to a population resistant to the killer. One result is that cotton farmers in the very States of the US where resistance to Darwinian ideas is at its strongest are struggling every season with the consequences of evolution at work in their own fields.

In hospitals, bacteria that cause diseases are increasingly resistant to drugs such as penicillin, for the same reason. The drugs kill all the susceptible bacteria, but, by definition, the survivors of an attack are the ones who are not susceptible to the drug. The more you kill susceptible bacteria, the more opportunities you give the others to spread – and bacteria breed far more quickly than human beings. The surprise, to an evolutionary biologist, is not that after half a century of use penicillin is losing its effectiveness, but rather that it retains any effectiveness at all.

But *why* are some people so hostile to the idea of

evolution? Perhaps it isn't the idea they are hostile to at all. In one of the most telling passages of his book, Weiner describes how one of the scientists involved in the research told of a long plane flight during which he got chatting to his neighbour, and described in detail what his work was all about.

'The whole time on that plane, my fellow passenger was getting more and more excited. "What a neat idea! What a neat idea!" Finally, as the plane was landing, I told him that the idea is called evolution. He turned purple.'

Just maybe, some of the people who turn purple at the mention of the word 'evolution' will read this book, and find out what a neat idea it really is, and how it explains what is going on in the world around us, not just among the finches of the Galápagos islands but in fields sprayed with insecticides and hospital wards where doctors find diminishing returns from the latest wonder drug. *The Beak of the Finch* is not the best place to learn about evolution if you have an open mind – Richard Dawkins' books still fit that brief. But it is the ideal book to recommend to any doubter who asks 'where's the evidence for evolution', and it is an entertaining insight into one of the most important pieces of biological research of the past twenty years. Those who already think they know a thing or two about evolution will read it with pleasure, coming away with a feeling that yes, of course, this is the way the world works.

Index

CD = Charles Darwin
n after page number = note
Shortened forms are used for long book titles